Books published in this series focus on the theory and computational foundations, advanced methodologies and practical applications of machine learning, ideally combining mathematically rigorous treatments of a contemporary topics in machine learning with specific illustrations in relevant algorithm designs and demonstrations in real-world applications. The intended readership includes research students and researchers in computer science, computer engineering, electrical engineering, data science, and related areas seeking a convenient medium to track the progresses made in the foundations, methodologies, and applications of machine learning.

Topics considered include all areas of machine learning, including but not limited to:

- Decision tree
- Artificial neural networks
- Kernel learning
- Bayesian learning
- Ensemble methods
- Dimension reduction and metric learning
- Reinforcement learning
- Meta learning and learning to learn
- Imitation learning
- Computational learning theory
- Probabilistic graphical models
- Transfer learning
- Multi-view and multi-task learning
- Agents and Multi-Agent Systems
- Graph neural networks
- Generative adversarial networks
- Federated learning
- Large Language Models
- Multimodal Learning
- Transformer/Diffusion Models
- Generative Artificial Intelligence
- Bio-inspired Learning Models
- Embodied AI
- Explainable AI and Ethics

This series includes monographs, introductory and advanced textbooks, and state-of-the-art collections. Furthermore, it supports Open Access publication mode.

Machine Learning: Foundations, Methodologies, and Applications

Yanghui Rao • Qing Li

Topic Modeling

Advanced Techniques and Applications

Springer

Yanghui Rao
School of Computer Science and Engineering
Sun Yat-sen University
Guangzhou, China

Qing Li
Department of Computing
The Hong Kong Polytechnic University
Hong Kong, China

ISSN 2730-9908 ISSN 2730-9916 (electronic)
Machine Learning: Foundations, Methodologies, and Applications
ISBN 978-981-96-8852-4 ISBN 978-981-96-8853-1 (eBook)
https://doi.org/10.1007/978-981-96-8853-1

This Springer imprint is published by the registered company Springer Nature Singapore Pte Ltd.
The registered company address is: 152 Beach Road, #21-01/04 Gateway East, Singapore 189721, Singapore

Preface

As a well-known text mining tool, topic modeling can effectively discover the latent semantic structure of text data. Extracting topics from such data is also one of the core issues in natural language processing. Topic information obtained from the processing on text data can be applied by researchers of both information science and social science areas. For instance, social science disciplines can use the topic extracted from text streams as a foundation to draw valuable analytical conclusions, including sentiment drift, public opinion changes, and psychological evolution.

Although topic modeling has largely improved and progressed greatly over the past three decades, it is often deemed as a hard problem. One of the reasons is due to the massive computational resource in parameter inference, especially for probabilistic topic models equipped with Monte Carlo sampling. To reduce computational cost, the Variational Auto-Encoder (VAE) and the Generative Adversarial Network (GAN) have been adopted to develop various neural topic models. Unfortunately, there are very few such methods designed to model the temporal aspect. Besides, many newer topic models still suffer from noise pollution in social media data. In this book, a number of approaches proposed to address these two limitations are introduced. First, typical lifelong topic models can accumulate knowledge learned from past for the future task, but the permutation of topics extracted by these methods may change over time, resulting in a semantic misalignment between the topic representations of document chunks across the stream. Such a misalignment deteriorates the model performances of various downstream tasks, including online document classification and dynamic information retrieval at the topic level. To address this issue, we exploit the Non-negative Matrix Tri-Factorization (NMTF) with a distributed parallel algorithm for lifelong topic modeling. Second, the issue of coherent topic modeling is also pertinent given the noise present in and large scale of social media data sets. A piece of message in social media platforms may include only a few words, rendering the lack of significant context. Models directly applied to these types of text often suffer from the feature sparsity problem, leading to undesirable results. To enhance coherence, we propose a series of topic models based on the high-quality embedding space. Third, a high coherence score of topics may be insufficient to indicate that they are "good" topics for content analysis. For example, the movie reviews are often categorized by their sentiment orientations (positive or negative), and we are interested in what a "positive/negative" topic is like. To achieve this, we develop supervised neural topic models by associating the discovered topics with human-defined labels automatically.

In the application scenario of emotion detection, public emotions always vary from one topic to another, and topics trigger public emotions. Therefore, topic discovery and emotion detection are closely related such that jointly modeling topics and emotions is an appropriate way to conduct these tasks. In this book, the effect of topics in emotion detection and related fields is also explored. To facilitate reproduction, we release the datasets and source codes related to the contemporary methods described in this book to https://github.com/YRaoGroup.

Guangzhou, China
Hong Kong SAR, China
April 2025

Yanghui Rao
Qing Li

Acknowledgments Dr. Celine Chang was the first editor from Springer Nature who encouraged us enthusiastically to come up with this timely book. Sudha Ramachandran as the project coordinator (book) has been patiently yet persistently reminding us from time to time on the progress of developing this book on time.

We owe our heartfelt gratitude to the members of our research group who have offered much help to the drafting of the book: Jiaxing Yan (Chapter 1 and Section 3.1), Ruihao Chen (Sections 2.1 and 3.2.1), Jiyuan Liu (Sections 2.2 and 3.2.2, Chapter 5), Yuyin Lu (Section 3.2.3), Zhiyu Shen (Section 4.1), and Yunhe Pang (Section 4.2); in particular, the contents of Sections 2.1.4 and 2.2.4 were mainly contributed by Ruihao Chen and Zhihong Zhang, respectively. We acknowledge the help of Jiyuan Liu, Jiaxing Yan, Zhiyu Shen, Yunhe Pang, Shanrui Xie, Juntian Huang, and Shizhen Li for their improvements to Sections 2 to 4; in addition, we also thank Hegang Chen, Zhiqi Lei, Zusheng Zhang, Lihui Lin, Ziye Chen, Jiachun Feng, Jianhui Pang, Jiemin Wu, Minghui Huang, Jiahai Wang, Hanjiang Lai, Jianxing Yu, Jian Yin, Haoran Xie, Fu Lee Wang, Raymond Y. K. Lau, Xizhao Wang, and Tak-Lam Wong for their valuable inputs and help.

We would like to thank our families. This project took valuable time away from them during the last 6 months. We appreciate their kind understanding and patience, as well as their great support during the writing of this book.

Competing Interests The authors have no competing interests to declare that are relevant to the content of this manuscript.

April 2025

Yanghui Rao
Qing Li

Contents

Acronyms

We provide a reference describing the abbreviations used throughout this book.

KDD	Knowledge Discovery in Databases
NLP	Natural Language Processing
LDA	Latent Dirichlet Allocation
VAE	Variational Auto-Encoder
GAN	Generative Adversarial Network
LLM	Large Language Model
KB	Knowledge Base
OT	Optimal Transport
NADE	Neural Autoregressive Distribution Estimator
ECR	Embedding Clustering Regularization
NMCM	Neural Mixed Counting Model
NVI	Neural Variational Inference
SLTM	Siamese Labeled Topic Model
NB-NTM	Negative Binomial-Neural Topic Model
GNB-NTM	Gamma Negative Binomial-Neural Topic Model
TSAM	Topic Self-Adaptive Model
TSNTM	Tree-Structured Neural Topic Model
LCNTM	Label-Conditioned Neural Topic Model
ETM	Embedding Topic Model
ECRTM	Embedding Clustering Regularization Topic Model
SBP	Stick-Breaking Process
CRP	Chinese Restaurant Process
HDP	Hierarchical Dirichlet Process
FNTM	Forest-Structured Neural Topic Model
GMHTM	Gaussian Mixture Hierarchical Topic Model
NMF	Non-negative Matrix Factorization
NMTF	Non-negative Matrix Tri-Factorization
NSEM	Nonlinear Structural Equation Model
FANTM	Forest-structured Adversarial Neural Topic Model

LNCHTM	Lifelong Nonparametric Clustering-based Hierarchical Topic Model
LANTM	Label Aligned Neural Topic Model
AP-HTM	Anchor word clustering and Path guided Hierarchical Topic Model
BoxTM	Box embedding-based Topic Model
UGTE	User Group-based Topic-Emotion model
WLTM	Weighted Labeled Topic Model
XETM	X-term Emotion-Topic Model
TADC	Topic-Aware Dynamic Convolutional neural network

Chapter 1
Introduction

Abstract Topic discovery from text data is one of the main subfields of Knowledge Discovery in Databases (KDDs); both aim to extract meaningful patterns from large datasets. KDD typically involves steps such as data cleaning and transformation, while topic discovery (aka topic modeling) methods similarly require text data preprocessing and semantic embedding. Topic modeling also has deep intersections with Natural Language Processing (NLP), and in recent years, methods utilizing pretrained or large language models to discover topics have been proposed. With the rapid development of KDD and NLP, topic modeling has seen significant achievements over the past three decades. Unfortunately, there remains a scarcity of topic models that effectively exploit the temporal aspect, especially the semantic evolution of document streams. Moreover, many contemporary topic models continue to grapple with the issue of noise contamination, particularly in social media data. In this chapter, we will first introduce the mainstream methods for topic modeling, i.e., probabilistic graphical models, nonnegative matrix factorization, and neural networks. Then, we will briefly describe hierarchical and lifelong topic modeling, in addition to the frequently used datasets and evaluation metrics. Finally, we will summarize the challenges in modern topic modeling and topic-level emotion detection from massive short texts.

Topic modeling is a widely adopted strategy in text mining systems, and it is also a fundamental module in natural language processing [3]. The target of topic modeling is to generate a set of coherent topics, and each topic consists of representative words. Usually, the topic can be represented by a distribution over the vocabulary, with weights denoting the relevancy of words to each topic. An additional representation called the document-topic distribution, with weights denoting the relevancy of topics to each document, can also be generated through the procedure of topic modeling.

The mainstream methods for topic modeling include probabilistic graphical models, Nonnegative Matrix Factorization (NMF), and neural networks. The probabilistic graphical model is a graphical representation of the conditional dependencies between random variables, commonly used to model complex data distributions.

Y. Rao, Q. Li, *Topic Modeling*, Machine Learning: Foundations, Methodologies, and Applications, https://doi.org/10.1007/978-981-96-8853-1_1

Latent Dirichlet Allocation (LDA) [2] is the most famous probability graph model in topic modeling, also known as a *probabilistic topic model*. LDA assumes that each document is composed of a mixture of topics, and each topic is a probability distribution over words. LDA estimates these implicit topic distributions and topic-word distributions through Bayesian inference. The inference is usually implemented through Monte Carlo sampling [1] or variational inference. NMF [16] is a matrix factorization technique that decomposes a nonnegative matrix into the product of two nonnegative matrices. In the context of the topic model, the original matrix can be seen as a document-word matrix, where rows correspond to documents and columns correspond to words in the vocabulary. Through NMF, two low-dimensional matrices can be obtained: one representing the document-topic matrix and the other representing the topic-word matrix. The advantage of NMF is that its results have intuitive interpretability and do not require complex probability inference like LDA. Recently, a popular method for topic modeling using neural networks, also known as a *neural topic model*, has attracted increasing attention. Neural topic models often adopt the framework of Variational Auto-Encoder (VAE) [14], in addition to that specifically designed for text, such as the Neural Autoregressive Distribution Estimator (NADE) [15] and the Recurrent Neural Network (RNN). These models are capable of capturing complex patterns in textual data, including long-term dependencies and semantic information. Neural topic models typically use pretrained language models such as BERT to better initialize weights and improve topic quality. In addition, neural network methods can learn richer and more abstract topic representations but often require a large amount of computing resources. Some clustering methods, such as k-means and the hierarchical DBSCAN, are also applied in topic modeling for efficient generation of topics regardless of the complex document generation procedure.

Different from the aforementioned methods, further scenarios such as hierarchical topic modeling, which constructs hierarchical structures for topics, dynamic topic modeling, which traces the evolution of topics over time, and lifelong topic modeling, which stores knowledge for text stream topic mining, are also being explored. Figure 1.1 illustrates a sample topic tree generated by hierarchical topic models. These topic models have been applied in many areas such as text analysis and content recommendation. The frequently used datasets include news reports (20News,[1] AGnews[2]), social media messages (Twitter,[3] Yahoo Answers[4]), reviews (Amazon Reviews,[5] IMDB[6]), scientific articles (Arxiv,[7] NIPS,[8] and DBLP[9]), and

[1] http://qwone.com/~jason/20Newsgroups

[2] http://groups.di.unipi.it/~gulli/AG_corpus_of_news_articles.html

[3] https://dumps.wikimedia.org/enwiki/latest

[4] https://github.com/LC-John/Yahoo-Answers-Topic-Classification-Dataset

[5] https://nijianmo.github.io/amazon/index.html

[6] https://ai.stanford.edu/~amaas/data/sentiment

[7] https://huggingface.co/datasets/arxiv-community/arxiv_dataset

[8] https://www.kaggle.com/datasets/benhamner/nips-papers

[9] https://github.com/JeloH/Dataset_DBLP

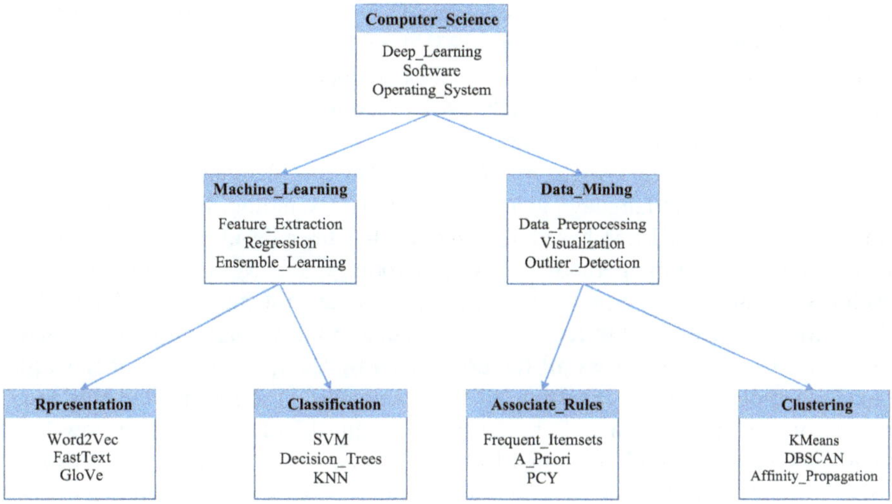

Fig. 1.1 A topic tree sample

so on. The benchmark datasets vary in length, size, and word distribution, which provides rich sources for text modeling and also poses challenges such as handling short texts, imbalanced text classification, and large datasets.

Traditional topic models are often unsupervised, which means that the evaluation is conducted without label information. Different from classical supervised metrics relying on the ground truth, the reliability and comprehensiveness of evaluation metrics for topic modeling are still under exploration. Firstly, the target of topic modeling is to obtain a set of topics, which are composed of representative words. For a single topic, we can evaluate its coherence by extracting the top words in it. Typical metrics for topic coherence include C_{Umass}, C_{UCI}, C_{npmi}, and C_v [23], which are variants by calculating the mutual information between two words. For the whole set evaluation, an intuitive perspective is that the topics need to be different. Therefore, the diversity can be estimated over the topic set by counting the redundant words. Two different metrics, Topic Uniqueness (TU) [20] and Topic Diversity (TD) [11], are widely applied by underestimating the scores of redundant words. With respect to hierarchical topic models, some similar metrics are designed for the parent-child topic relation evaluation. CLNPMI [9] is the coherence between parent and child topics by calculating the mutual information over the distinct words in two sets, and the overlap rate is used by counting the redundancy of top words between parent and child topics. Secondly, the document-topic distribution can be obtained by most topic models, and it reveals the quality of topic modeling to some degree. In this case, text clustering and classification are often applied as the downstream tasks. For better comprehension, some human-involved tasks are also proposed for explicit coherence evaluation. The word intrusion and topic intrusion tasks [6] can express the coherence between the top words in a topic and the top-related topics for a document. They can also provide references for the aforementioned automated metrics [23]. Finally, in the early research on topic modeling, researchers tended to treat it as a document generation procedure. Therefore, perplexity, which is widely used in language models, is still applied in the evaluation of various topic models.

Recently, the Large Language Model (LLM) has demonstrated remarkable capabilities in natural language generation and understanding. In light of this consideration, a series of LLM-based topic models has been developed [5, 12, 19, 21]. However, most existing methods involve prompting the LLM to generate topics for each document in the corpus, which may lead to several limitations. First, since the LLM is mainly asked to focus on one document individually, it may be unable to cover all the topics across all the documents in a corpus [12]. In other words, modeling topics requires inputting all the documents in a corpus, and this long-range context information may be forgotten by the LLM and result in low-level topic quality. Second, although the LLM excels at capturing local context, it usually struggles with long documents with multiple interrelated topics, which may evolve or shift throughout the text.

Through the summarization and analysis of both classic and popular topic models, we have drawn the following limitations. First, there remains a scarcity of topic models that effectively exploit the temporal aspect, especially the semantic evolution of document streams. Second, many contemporary topic models continue to grapple with the issue of noise contamination, particularly in social media data.

Lifelong topic models are developed to capture the dynamic semantics of a stream of documents by accumulating knowledge learned from experience for future tasks, which can be categorized into methods based on probabilistic graphical models [10], NMF [8, 22], and neural networks [13, 26]. However, the sequence of topics extracted by these methods may shift over time, leading to semantic misalignment between the topic representations across document streams. Such misalignment can degrade the performances of various downstream tasks, including online document classification and dynamic information retrieval at the topic level.

Additionally, the challenge of coherent topic modeling is particularly relevant due to the noise and large scale of social media datasets. Messages on social media platforms often consist of only a few words, resulting in a lack of significant context. Models applied directly to this type of text frequently encounter the problem of feature sparsity, which can yield unsatisfactory outcomes. To enhance coherence, most of the contemporary topic models turn to utilizing embeddings [11]. By manipulating topic embeddings and word embeddings on the embedding space, we can achieve stronger power of discovering high-quality topics. For example, by formalizing an optimal transport problem between topic embeddings and word embeddings, Wu et al. [24] proposed the Embedding Clustering Regularization (ECR) to avoid topic collapsing. Multiple geometric embedding techniques, like hyperbolic embedding [25] and box embedding [18], are also considered. In addition to exploring embeddings for coherent topic modeling, researchers further attempt to build fine-grained dependencies between topics to enhance interpretability and better model the documents. To this end, a hierarchical topic model via anchor word clustering and path guidance was developed [17]. Considering that metadata provides additional information about the document [4], we also proposed a supervised topic model which can greatly enhance the alignment between labels and topics while maintaining good topic quality [7].

In the context of emotion detection, public emotions are known to fluctuate across different topics, and topics can evoke public emotion. Thus, there is a strong interconnection between topic discovery and emotion detection. Jointly modeling topics and emotions is a suitable strategy for these tasks. Unfortunately, traditional topic-emotion modeling approaches often fail to effectively capture user characteristics in the emotion detection process, which prevents them from discovering user-specific emotional tendencies. Besides, detecting emotions from massive short texts remains challenging due to the feature sparsity of short texts and the high computational complexity of topic-level emotion detection models.

References

1. Blei DM (2012) Probabilistic topic models. Commun ACM 55(4):77–84
2. Blei DM, Ng AY, Jordan MI (2003) Latent Dirichlet allocation. J Mach Learn Res 3:993–1022
3. Branavan SRK, Chen H, Eisenstein J, Barzilay R (2009) Learning document-level semantic properties from free-text annotations. J Artif Intell Res 34:569–603
4. Card D, Tan C, Smith NA (2018) Neural models for documents with metadata. In: Proceedings of the 56th annual meeting of the association for computational linguistics, pp 2031–2040
5. Çelikten T, Onan A (2025) Topic modeling through rank-based aggregation and LLMs: an approach for AI and human-generated scientific texts. Knowl-Based Syst 314:113219
6. Chang J, Gerrish S, Wang C, Boyd-Graber J, Blei D (2009) Reading tea leaves: how humans interpret topic models. In: Advances in neural information processing systems, pp 288–296
7. Chen R, Chen H, Lu Y, Rao Y, Zhu C (2005) Supervised neural topic modeling with label alignment. Trans Assoc Comput Linguist 13:249–263
8. Chen Y, Wu J, Lin J, Liu R, Zhang H, Ye Z (2020) Affinity regularized non-negative matrix factorization for lifelong topic modeling. IEEE Trans Knowl Data Eng 32(7):1249–1262
9. Chen Z, Ding C, Zhang Z, Rao Y, Xie H (2021) Tree-structured topic modeling with nonparametric neural variational inference. In: Proceedings of the 59th annual meeting of the association for computational linguistics and the 11th international joint conference on natural language processing, pp 2343–2353
10. Chen Z, Liu B (2014) Topic modeling using topics from many domains, lifelong learning and big data. In: Proceedings of the 31st international conference on machine learning, pp 703–711
11. Dieng AB, Ruiz FJ, Blei DM (2020) Topic modeling in embedding spaces. Trans Assoc Comput Linguist 8:439–453
12. Doi T, Isonuma M, Yanaka H (2024) Topic modeling for short texts with large language models. In: Proceedings of the 62nd annual meeting of the association for computational linguistics (student research workshop), pp 21–33
13. Gupta P, Chaudhary Y, Runkler TA, Schütze H (2020) Neural topic modeling with continual lifelong learning. In: Proceedings of the 37th international conference on machine learning, pp 3907–3917
14. Kingma DP, Welling M (2013) Auto-encoding variational Bayes. In: Proceedings of the 2nd international conference on learning representations
15. Lauly S, Zheng Y, Larochelle H et al (2017) Document neural autoregressive distribution estimation. J Mach Learn Res 18(113):1–24

16. Lee D, Seung HS (2000) Algorithms for non-negative matrix factorization. In: Advances in neural information processing systems, pp 556–562
17. Liu J, Chen H, Zhu C, Rao Y (2024) Unsupervised hierarchical topic modeling via anchor word clustering and path guidance. In: Findings of the association for computational linguistics: EMNLP, pp 7505–7517
18. Lu Y, Chen H, Mao P, Rao Y, Xie H, Wang FL, Li Q (2024) Self-supervised topic taxonomy discovery in the box embedding space. Trans Assoc Comput Linguist 12:1401–1416
19. Mu Y, Dong C, Bontcheva K, Song X (2014) Large language models offer an alternative to the traditional approach of topic modelling. In: Proceedings of the 2024 joint international conference on computational linguistics, language resources and evaluation, pp 10160–10171
20. Nan F, Ding R, Nallapati R, Xiang B (2019) Topic modeling with Wasserstein autoencoders. In: Proceedings of the 57th annual meeting of the association for computational linguistics, pp 6345–6381
21. Pham C, Hoyle AM, Sun S, Resnik P, Iyyer M (2024) TopicGPT: a prompt-based topic modeling framework. In: Proceedings of the 2024 conference of the North American chapter of the association for computational linguistics: human language technologies, pp 2956–2984
22. Qin X, Lu Y, Chen Y, Rao Y (2011) Lifelong learning of topics and domain-specific word embeddings. In: Findings of the association for computational linguistics, pp 2294–2309
23. Röder M, Both A, Hinneburg A (2015) Exploring the space of topic coherence measures. In: Proceedings of the 8th ACM international conference on web search and data mining, pp 399–408
24. Wu X, Dong X, Nguyen TT, Luu AT (2023) Effective neural topic modeling with embedding clustering regularization. In: Proceedings of the 40th international conference on machine learning, pp 37335–37357 (2023)
25. Xu Y, Wang D, Chen B, Lu R, Duan Z, Zhou M (2022) HyperMiner: topic taxonomy mining with hyperbolic embedding. In: Advances in neural information processing systems
26. Zhang X, Rao Y, Li Q (2022) Lifelong topic modeling with knowledge-enhanced adversarial network. World Wide Web 25:219–238

Chapter 2
Classical Topic Models

Abstract Topic modeling aims to discover topics in text data, where each topic can be represented by a distribution over the vocabulary, with weights denoting the relevancy of words to the topic. By selecting a small number of words with the largest weights associated with every topic (aka top words), we can obtain word clusters that reflect the latent semantics of topics. For instance, a topic on computer hardware could be captured by a cluster of words such as "scsi," "bios," "drives," "controller," and "bus." In classical topic models, one stream of research focuses on mining diverse topics. The corresponding methods assume that topics are independent of each other. The other researchers concentrate on revealing the hierarchical relationships among topics, which are often observed in web pages and academic repositories. This chapter will summarize the classical topic modeling methods from the following two categories: flat topic models and hierarchical topic models. For the former, we introduce the flat topic models that utilize Siamese neural networks, neural mixed counting models, parallel Gamma-Gamma-Poisson processes, and variational auto-encoders, respectively. For the latter, we describe the hierarchical topic models that organize topics into various structures, including a tree, a forest, and more flexible graphs.

2.1 Flat Topic Models

One of the main classical methods for topic discovery is flat topic models, in which the dependencies among topics are not modeled. This setting simplifies the task and is therefore widely adopted. Here, we first introduce our works entitled "Siamese Network-Based Supervised Topic Modeling" [33], "Neural Mixed Counting Models for Dispersed Topic Discovery" [85], and "Copula Guided Parallel Gibbs Sampling for Nonparametric and Coherent Topic Discovery" [47]. Then, we present a preliminary study on incorporating document labels into existing flat topic models, which sheds light on our future work in this area.

Y. Rao, Q. Li, *Topic Modeling*, Machine Learning: Foundations, Methodologies, and Applications, https://doi.org/10.1007/978-981-96-8853-1_2

2.1.1 Siamese Labeled Topic Model

Traditional topic modeling approaches such as Latent Dirichlet Allocation (LDA) [5] are unsupervised and fail to incorporate labels or response variables during the learning process. This limitation prevents them from discovering label-specific topics, which are essential for multiple applications like aspect-based sentiment analysis. To address this issue, several supervised topic models have been developed. In the supervised LDA (sLDA) [3], each document was paired with a response variable which obeys the Gaussian distribution. The supervised Neural Topic Model (sNTM) [11] leverages two hidden layers to generate the n-gram topic and document-topic representations. Unfortunately, sNTM treats labels and topics separately, reducing the influence of label information on topic generation. To this end, we propose a Siamese Labeled Topic Model (SLTM), leveraging Siamese neural networks to achieve label-specific topic discovery and label prediction in a unified framework.

2.1.1.1 Introduction

Most early topic modeling approaches—like probabilistic Latent Semantic Analysis (pLSA) [29] and LDA [5]—are entirely unsupervised: they learn topics purely from word co-occurrence patterns and do not use any document labels. As a result, they cannot discover topics that are specific to particular labels, even though such "label-specific" topics are crucial for applications like personality psychology, aspect-level sentiment analysis, and cross-domain sentiment classification. To remedy this, a family of supervised topic models has been proposed. The sLDA [3] ties each document to a single real-valued response, while the multi-class sLDA (sLDAc) [81] and the Labeled LDA (LLDA) [67] extend this framework to categorical labels for classification tasks. Later, the sNTM [11] combines neural networks with topic modeling ideas to handle both classification and regression. However, in sNTM, the label information still has only a weak influence on topic formation, since it models the document-topic and label-topic relationships largely independently.

The above limitation motivates us to develop a supervised topic model which can jointly model documents and labels. Particularly, we propose the SLTM to exploit the information of documents and labels based on the Siamese network, where weight matrices in SLTM represent conditional distributions. Therefore, by constraining weight matrices during the learning procedure, SLTM can follow probabilistic characteristics of topic models strictly. Compared to previous supervised topic models, the main advantages of our SLTM are summarized as follows. First, SLTM can generate more coherent label-specific topics than others. This is because the supervision of labels is incorporated into topic modeling for SLTM. On the other hand, the mapping of topics to labels is unconstrained for most existing supervised topic models, which renders many coherent topics being generated outside labels. Second, strengths of neural networks are incorporated into SLTM to bootstrap its

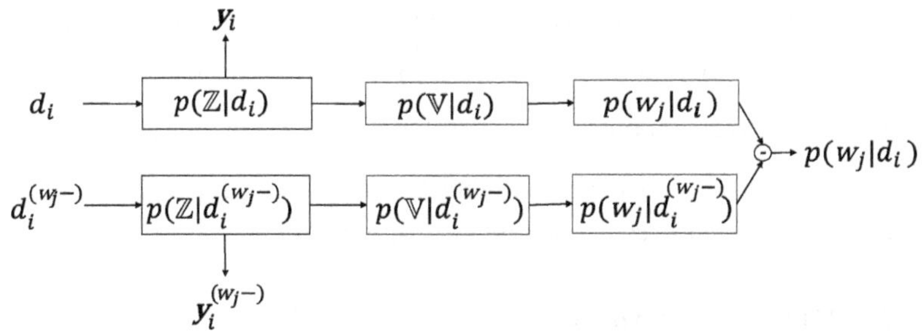

Fig. 2.1 The word generation process of SLTM, where $\mathbb{Z}$ and $\mathbb{V}$ denote the topic collection and vocabulary, respectively

inference power on label prediction. Third, each word can be mapped to a topical embedding space and represented by a word embedding after generating label-specific topics.

2.1.1.2 Methodology

The word generation process of SLTM is presented in Fig. 2.1. Specifically, we define a primary network that encodes the actual documents d_i into document-topic distributions $\boldsymbol{\theta}_i$ and predicts the labels y_i. Denote the number of topics, number of documents, and vocabulary size as K, M, and V, respectively. The primary network is parameterized by three weight matrices: $\Theta = (\boldsymbol{\theta}_1, \ldots, \boldsymbol{\theta}_M) \in \mathbb{R}^{K\times M}$, the document-topic distributions; $\Phi = (\boldsymbol{\phi}_1, \ldots, \boldsymbol{\phi}_K) \in \mathbb{R}^{V\times K}$, the topic-word distributions; and $\mathbf{W}_{tl} \in \mathbb{R}^{N_{label}\times K}$, the topic-label mappings, where N_{label} is the number of labels. $\boldsymbol{\theta}_i$ is obtained by $\Theta \boldsymbol{d}_i$, where $\boldsymbol{d}_i$ is the indicator vector of d_i within the corpus. Then, we compute the probability of each word conditioned on d_i by $\left(p(w_j|d_i)\right)_{j=1}^{V} = \Phi\boldsymbol{\theta}_i$ and compute the likelihood loss. To incorporate label information, we predict the label y_i by $\mathbf{W}_{tl}\boldsymbol{\theta}_i$ and compute the label loss.

In addition to the primary network, we introduce a contrastive network to further distinguish documents with or without w_j. For each word w_j, it encodes a negative sample document $d_i^{(w_j-)}$, which is randomly selected and excludes w_j. Following the same mechanism mentioned above, we compute the conditional word probabilities and predict $y_i^{(w_j-)}$. Because the document-topic distributions and the topic-word distributions of a corpus are fixed, Θ, Φ, and $\mathbf{W}_{tl}$ are shared among two networks. These two networks are twin networks, and thus the proposed model is essentially the Siamese network.

In order to discover coherent topics, we adopt the following likelihood loss:

$$loss\left(d_i, d_i^{(w_j-)}\right) = \left|p(w_j|d_i) - \hat{p}(w_j|d_i) + \hat{p}\left(w_j|d_i^{(w_j-)}\right)\right|, \tag{2.1}$$

which encourages the predicted word probabilities $\hat{p}(w_j|d_i)$ to match the observed frequencies while minimizing $\hat{p}\left(w_j|d_i^{(w_j-)}\right)$ for negative samples. In addition, to predict labels, we compute the label loss $loss\left(y_i, y_i^{(w_j-)}\right) = loss(y_i) + loss\left(y_i^{(w_j-)}\right)$ according to different label types (like categorical or real-valued labels). Finally, the total loss is a weighted combination:

$$loss(SLTM) = \alpha \times loss\left(d_i, d_i^{(w_j-)}\right) + (1-\alpha) \times loss\left(y_i, y_i^{(w_j-)}\right). \quad (2.2)$$

2.1.1.3 Experimental Results

To validate the effectiveness of SLTM, we evaluate it on two real-world datasets, ISEAR and YouTube, which contain emotion categories and sentiment strength, respectively. The coherence scores (measured by NPMI) indicate that SLTM can discover more coherent topics than both unsupervised topic models and supervised methods. It is also interesting to observe that supervised baseline models perform worse than pLSA and LDA for most cases, which validates that it is challenging to trade off label-specific word distributions with document-specific label distributions.

In qualitative analysis, we show that SLTM performs better than baseline models in label-specific topic discovery, although all models can learn meaningful topics. For example, two words "happy" and "joy" that are strongly related to the label of "joy" are identified by SLTM with large probabilities. Similar results can be observed in other labels; thus topics discovered by our model are more convenient to be understood than others.

To further evaluate the interpretability of topics extracted from SLTM, we obtain topic embeddings by $emb(z_k) = \mathbf{W}_1[k, :]$. Then, we map $emb(z_k)$ to a two-dimensional space via Principal Component Analysis (PCA). Figure 2.2 presents the

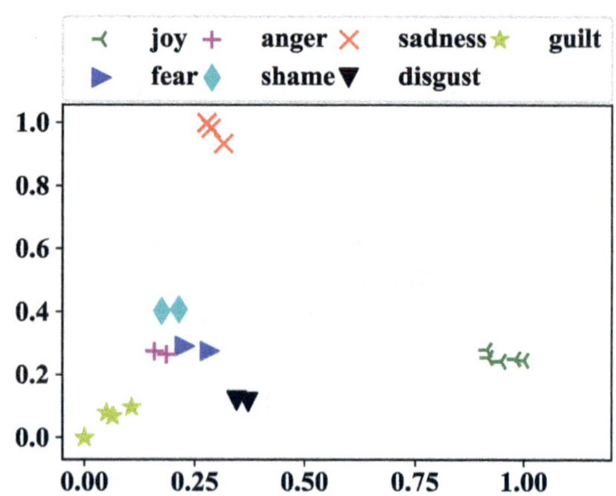

Fig. 2.2 Scatter plot of topics identified by SLTM on ISEAR, where each point indicates a topic

scatter plot of topics generated by SLTM on the ISEAR dataset. The visualization result indicates that topics corresponding to the same label are closer than those of different labels. Furthermore, the distance between topics on correlated labels such as "fear" and "anger" is closer than that of topics on "joy" and other labels.

2.1.2 Neural Mixed Counting Models

In the setting of topic modeling, documents are often featured by their bag-of-word vectors. Such count-valued representations are over-dispersed, which means the variance is larger than the mean. Mixed counting models [76, 96] that use the negative binomial distribution as the prior are proven to work well on over-dispersed data, and they have attracted much attention in the field of topic modeling. However, they depend on Monte Carlo sampling to conduct parameter inference, which is quite time-consuming. In the meantime, the Neural Variational Inference (NVI) [57] is developed and successfully applied to topic models. In order to combine the superiority of mixed counting models on over-dispersed data and the fast inference speed of NVI, we propose two efficient Neural Mixed Counting Models (NMCMs), Negative Binomial-Neural Topic Model (NB-NTM) and Gamma Negative Binomial-Neural Topic Model (GNB-NTM), which are based on the NB process [76] and the GNB process [96], respectively. The framework of NB-NTM is shown in Fig. 2.3, and the framework of GNB-NTM is shown in Fig. 2.4, where K is the number of topics which is set manually, r is the component weight of G, i.e., the topic measure at the corpus level, and $G = \sum_{k=1}^{K} r_k \delta_{\pi_k}$.

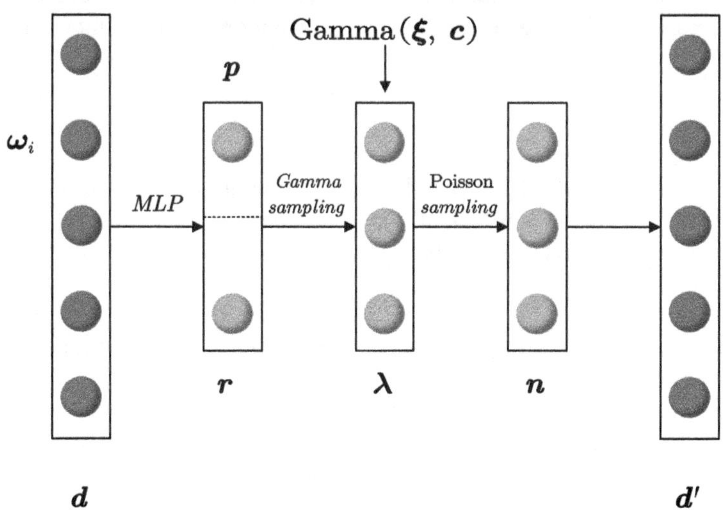

Fig. 2.3 Framework of NB-NTM

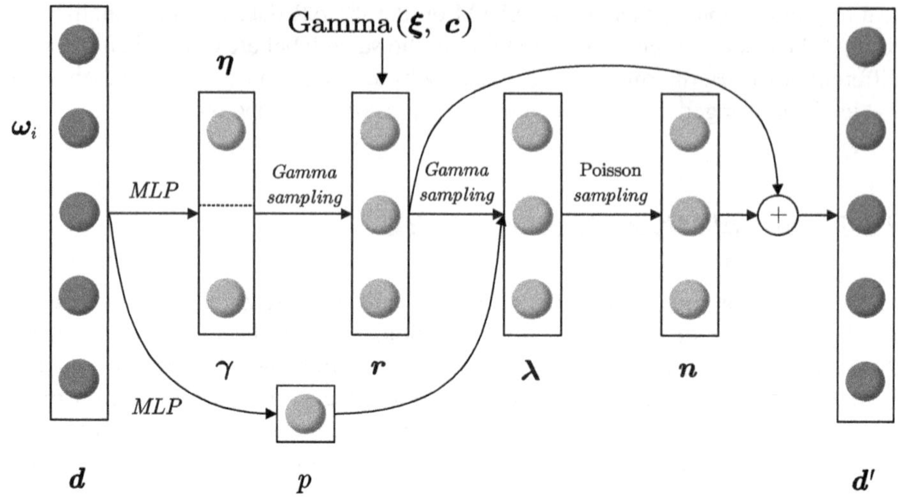

Fig. 2.4 Framework of GNB-NTM

$\boldsymbol{\lambda}$ represents the weights of topics at the document level, which can be used to estimate the topic measure on d by $\Lambda = \sum_{k=1}^{K} \lambda_k \delta_{\pi_k}$. In the above, λ_k denotes the kth component of $\boldsymbol{\lambda}$. $Gamma(\boldsymbol{\xi}, \boldsymbol{c})$ is the Gamma prior for $\boldsymbol{\lambda}$. Besides, $\boldsymbol{n}$ is the component weight of Π that represents a Poisson process at the word level, and $\Pi = \sum_{k=1}^{K} n_k \delta_{\pi_k}$. For GNB-NTM, γ and η are the parameters of the first-level Gamma process, and p is the scale parameter of the second-level Gamma process. As a novel framework for dispersed topic discovery, NMCMs including NB-NTM and GNB-NTM integrate neural networks with flexible count-based generative models to parameterize flexible probability distributions for topic modeling.

2.1.2.1 Introduction

Mixture modeling provides a principled way to assign data points—such as words in documents—to latent clusters or topics by drawing from random probability measures. Traditional approaches like the Hierarchical Dirichlet Process (HDP) capture dependencies across groups (e.g., documents and topics) but rely on complex indirect constructions (e.g., Chinese Restaurant Franchise and stick-breaking) and costly inference schemes (e.g., Monte Carlo sampling and variational methods), making them cumbersome to scale and implement.

An alternative class of "mixed counting" models based on completely random measures—the Negative Binomial (NB) and Gamma Negative Binomial (GNB) processes—naturally yields independent generative steps, enforce nonnegativity, and handle over-dispersed counts (i.e., the variance exceeds the mean). In fact, a normalized GNB process is equivalent to HDP, while also offering more flexible hierarchical dependencies. Yet, as datasets grow, both sampling-based and traditional variational inferences become computationally prohibitive or mathematically intractable for these richer count models.

Neural Variational Inference (NVI) frameworks—extending variational auto-encoders to discrete and continuous latent structures—have enabled fast and scalable topic models (aka neural topic models) that learn document-to-topic mappings via backpropagation (using reparameterization or variance-reduction techniques). Despite strong empirical performance, these neural topic models often sacrifice interpretability of their hidden variables and struggle to capture over-dispersion and hierarchical topic dependencies.

We propose two novel neural mixed counting models dubbed the Negative Binomial-Neural Topic Model (NB-NTM) and the Gamma Negative Binomial-Neural Topic Model (GNB-NTM) based on NB and GNB processes, respectively. The general motivation is to combine the advantages of NVI and mixed counting models. On the one hand, NVI-based models are fast and easy to estimate but hard to interpret. On the other hand, document modeling via mixed counting models is easy to interpret but difficult to infer. In our NB-NTM and GNB-NTM, we develop NVI algorithms to infer parameters by using the reparameterization of Gamma distribution and the Gaussian approximation of Poisson distribution. Extensive experiments on real-world datasets validate the effectiveness of our proposed models in perplexity, topic coherence, and dispersed topic learning. Furthermore, the proposed models can describe the hierarchical dependence of random probability measures and introduce nonnegative constraints, which renders the intermediate variables generated by our methods to have good interpretability.

2.1.2.2 Methodology

At the core of NB-NTM is a Negative Binomial (NB) distribution, which serves as the base for modeling over-dispersed word counts in documents. The NB distribution can be extended to a Gamma distribution and a Poisson distribution. Therefore, within the framework of Variational Auto-Encoder (VAE), we encode the document d to parameterize these two distributions using the bag-of-word representation $\boldsymbol{x}_d$:

$$\boldsymbol{r} = f_1(\boldsymbol{x}_d), \;\; \boldsymbol{p} = f_2(\boldsymbol{x}_d), \tag{2.3}$$

$$\boldsymbol{\lambda} \sim \text{Gamma}(\boldsymbol{r}, \boldsymbol{p}/(\boldsymbol{1} - \boldsymbol{p})), \;\; \boldsymbol{n} \sim \text{Poisson}(\boldsymbol{\lambda}). \tag{2.4}$$

To reconstruct d, we compute the following conditional probability over each word w_j using the latent variables $\boldsymbol{n}$:

$$p(w_j|\boldsymbol{\lambda}) = \frac{\exp\left(\sigma\left(\boldsymbol{n}^T \Phi^T \boldsymbol{w}_j + \boldsymbol{b}_j\right)\right)}{\sum_{j'=1}^{V} \exp\left(\sigma\left(\boldsymbol{n}^T \Phi^T \boldsymbol{w}_{j'} + \boldsymbol{b}_{j'}\right)\right)}, \tag{2.5}$$

where $\boldsymbol{w}_j$ is the indicator vector of word w_j within the vocabulary. $\Phi = (\boldsymbol{\phi}_1, \ldots, \boldsymbol{\phi}_K) \in \mathbb{R}^{V \times K}$ is the topic-word distributions. $\boldsymbol{b}$ is a learnable bias term, and σ is the sigmoid activation function. We present the parameter inference process of NB-NTM in Algorithm 1, in which the variational lower bound $\mathcal{L}$ is used to calculate gradients and model parameters are updated by Adam [39].

Algorithm 1: Parameter inference for NB-NTM

Input: Number of topics K, gamma priors $\boldsymbol{\xi}$ and $\boldsymbol{c}$, document set $\boldsymbol{D}$;
Output: Document-topic distribution $\boldsymbol{\theta}$, topic-word distribution $\boldsymbol{\phi}$.

repeat
  for *document* $\boldsymbol{d} \in \boldsymbol{D}$ **do**
    Compute gamma distribution parameters $\boldsymbol{r} = f_1(\boldsymbol{d})$, $\boldsymbol{p} = f_2(\boldsymbol{d})$;
    Compute the KL divergence between Gamma($\boldsymbol{r}, \boldsymbol{p}$) and Gamma($\boldsymbol{\xi}, \boldsymbol{c}$);
    for $k \in [1, K]$ **do**
      Sample the Poisson distribution parameter by $\lambda_k \sim \text{Gamma}(r_k, p_k/(1-p_k))$;
      Sample word numbers by $n_k \sim$ Poisson (λ_k);
    for $\omega_i \in \boldsymbol{d}$ **do**
      Compute log-likelihood $\log p(\boldsymbol{\omega}_i | \boldsymbol{\lambda})$;
    Compute variational lower bound $\mathcal{L}$;
    Update $f_1(\cdot)$, $f_2(\cdot)$, R, and $\boldsymbol{b}$;
until *convergence*;
for *document* $\boldsymbol{d} \in \boldsymbol{D}$ **do**
  Normalize $\boldsymbol{\lambda}$ to obtain θ_d;
Apply softmax to R in row to obtain $\boldsymbol{\phi}$.

Based on NB-NTM, we further propose GNB-NTM by assigning another Gamma process as a prior to the NB process. Therefore, the encoding is described as follows:

$$\boldsymbol{\gamma} = f_1(\boldsymbol{x}_d),\ \boldsymbol{\eta} = f_2(\boldsymbol{x}_d),\ \boldsymbol{r} \sim \text{Gamma}(\boldsymbol{\gamma}, \boldsymbol{\eta}), \tag{2.6}$$

$$p = f_3(\boldsymbol{x}_d),\ \boldsymbol{\lambda} \sim \text{Gamma}(\boldsymbol{r}, p/(1-p)),\ \boldsymbol{n} \sim \text{Poisson}(\boldsymbol{\lambda}). \tag{2.7}$$

Since an extra Gamma process is introduced over the existing Gamma process of NB-NTM, GNB-NTM is able to characterize the multilevel dependencies of random variables and thus more flexible and powerful.

To conduct NVI, we follow existing practices, using the reject-sampling method [59] for the Gamma distribution $x \sim \text{Gamma}(\alpha, \beta)$: $x = \frac{1}{\beta}\left(\alpha - \frac{1}{3}\right)\left(1 + \frac{\varepsilon}{\sqrt{9\alpha - 3}}\right)^3$, where $\varepsilon \sim \mathcal{N}(0, 1)$. For the Poisson distribution $x \sim \text{Poisson}(\lambda)$, we approximate it by $\mathcal{N}(\lambda, \lambda)$ based on the central limit theorem. The parameter inference for GNB-NTM is presented in Algorithm 2. We use the variational lower bound to calculate gradients and apply Adam to update parameters of GNB-NTM, which are the same as NB-NTM.

Algorithm 2: Parameter inference for GNB-NTM

Input: Number of topics K, gamma priors $\boldsymbol{\xi}$ and $\boldsymbol{c}$, document set $\boldsymbol{D}$;
Output: Document-topic distribution $\boldsymbol{\theta}$, topic-word distribution $\boldsymbol{\phi}$.

repeat
  for *document* $\boldsymbol{d} \in \boldsymbol{D}$ **do**
    Compute the 1st gamma distribution parameters $\boldsymbol{\gamma} = f_1(\boldsymbol{d})$, $\boldsymbol{\eta} = f_2(\boldsymbol{d})$;
    Compute the KL divergence between Gamma$(\boldsymbol{\gamma}, \boldsymbol{\eta})$ and Gamma$(\boldsymbol{\xi}, \boldsymbol{c})$;
    Compute the 2nd gamma distribution parameter $p = f_3(\boldsymbol{d})$;
    for $k \in [1, K]$ **do**
      Sample the 2nd gamma distribution parameter by $r_k \sim \text{Gamma}(\gamma_k, \eta_k)$;
      Sample the Poisson distribution parameter by $\lambda_k \sim \text{Gamma}(r_k, p/(1-p))$;
      Sample word numbers by $n_k \sim \text{Poisson}(\lambda_k)$;
    for $\omega_i \in \boldsymbol{d}$ **do**
      Compute log-likelihood $\log p(\boldsymbol{\omega}_i | \boldsymbol{r})$;
    Compute variational lower bound $\mathcal{L}$;
    Update $f_1(\cdot)$, $f_2(\cdot)$, $f_3(\cdot)$, R, and $\boldsymbol{b}$;
until *convergence*;
for *document* $\boldsymbol{d} \in \boldsymbol{D}$ **do**
  Normalize $\boldsymbol{\lambda}$ to obtain $\theta_{\boldsymbol{d}}$;
Apply softmax to R in row to obtain $\boldsymbol{\phi}$.

2.1.2.3 Experimental Results

We employ the following three datasets to evaluate the effectiveness of our models: Reuters,[1] 20News, and MXM song lyrics [54]. The Reuters dataset contains 7758 training documents and 3005 testing documents. The 20News corpus consists of 18,773 news articles under 20 categories. These news articles are divided into 11,268 training documents and 7505 testing documents. The 20 categories include sports, electronics, automotive, and so forth, and the number of documents under each category is almost the same. MXM is the official lyrics collection of the Million Song Dataset, which contains 210,519 training documents and 27,143 testing documents, respectively. By following [54], we use the originally provided vocabulary with 5000 words for MXM, while for Reuters and 20News, we use stemming, stop words filtering, and the 2000 most frequently occurred words as vocabularies. The statistics of these datasets are presented in Table 2.1.

To evaluate the quality of topics generated by our proposed models and other baseline methods, we use perplexity and topic coherence as evaluation criteria. Table 2.2 shows the perplexity and topic coherence of different models on the test datasets. We can observe that NB-NTM outperforms most baselines, and GNB-NTM performs the best in all cases. The results validate that the NB distribution can model over-dispersed documents well.

[1] https://www.nltk.org/book/ch02.html

Table 2.1 Statistics of the adopted datasets

Dataset	Reuters	20News	MXM
Training set size	7758	11,268	210,519
Testing set size	3005	7505	27,143
Label number	90	20	–
Vocabulary size	2000	2000	5000

Table 2.2 Perplexity and topic coherence results, where the latter is an average of three coherence scores by calculating 5, 10, and 15 top words for each topic

	Perplexity			Topic coherence		
Model	Reuters	20News	MXM	Reuters	20News	MXM
HDP	302.3	730.7	319.5	0.305	0.223	0.356
NVDM	224.9	855.0	252.3	0.133	0.138	0.109
NVLDA	578.6	1252.2	668.7	0.253	0.240	0.216
prodLDA	648.1	1267.2	852.9	0.332	0.329	0.313
GSM	266.2	963.5	330.5	0.192	0.211	0.177
iDocNDAE	202.8	844.6	294.7	0.130	0.151	0.222
NB-NTM	181.0	740.6	247.5	0.341	0.343	0.340
GNB-NTM	**146.7**	**602.8**	**216.8**	**0.377**	**0.375**	**0.427**

To further see how our proposed models perform on dispersed datasets, we first count the total number of words under each manually labeled category (i.e., topic) as the topic-word number distribution. Then, we run our NB-NTM and GNB-NTM on the testing set of 20News to get the corresponding values of $\boldsymbol{r}$. After normalization, we show the proportion of different topics obtained by NB-NTM and GNB-NTM at the corpus level. The results are shown in Fig. 2.5, where (a) is the topic-word number distribution generated by the manually labeled category that is treated as a topic, (b) is the proportion of 20 topics obtained by NB-NTM, and (c) is the proportion of 20 topics obtained by GNB-NTM.

From these results, we can observe that the proportion of topics obtained by NB-NTM is close to the topic-word number distribution. On the other hand, GNB-NTM obtains more dispersed proportions of topics than NB-NTM. These results suggest that GNB-NTM tends to allocate less but more important topics to the corpus, i.e., the topics generated by GNB-NTM are more discriminative. Since the document-topic distribution is not directly modeled and the Gaussian distribution samples are not nonnegative, the previous neural methods except GSM cannot obtain explainable intermediate variables.

2.1.3 *Topic Self-adaptive Model*

It is difficult for a parametric topic model to determine the optimal number of topics. Therefore, nonparametric topic models are developed, among which HDP [75] is

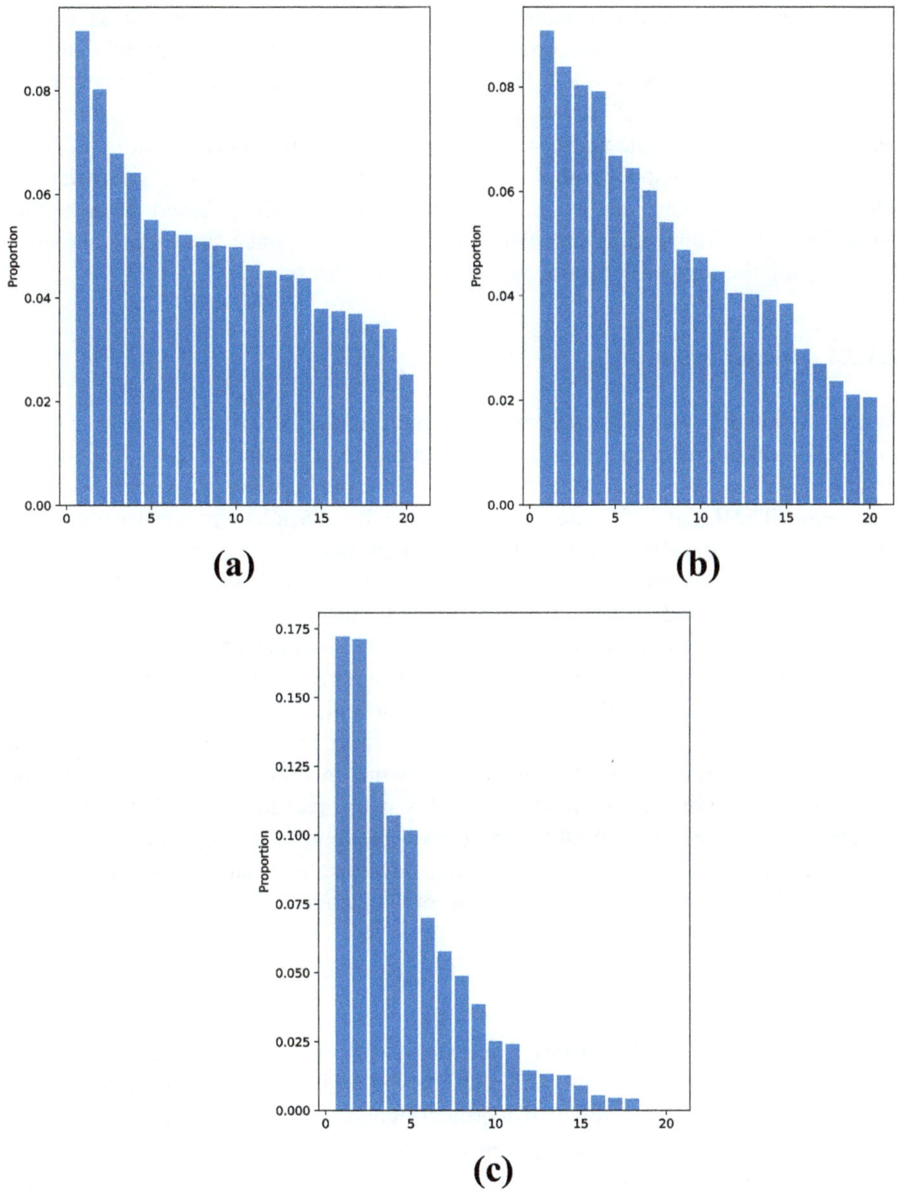

Fig. 2.5 Corpus-level topic proportions for (**a**) manual labels, (**b**) NB-NTM, and (**c**) GNB-NTM models using the testing set of 20News

the most well-known. Unfortunately, it requires massive computational resources when applying HDP to large corpora. To tackle this limitation, a parallel Gamma-Gamma-Poisson Process (G2PP) [16] was proposed for efficient topic generation. Although the parallel G2PP converges quite fast on large-scale corpora, this method requires a predefined topic number and thus loses the nonparametric feature of HDP. To this end, we propose a parallel nonparametric topic model, Topic Self-Adaptive Model (TSAM). Then, we extend our TSAM to TSAMcop based on a copula guided parallel Gibbs sampling algorithm, so as to enhance the topic coherence by modeling the topical dependencies within each phrase.

2.1.3.1 Introduction

In the classic LDA [5], documents have multinomial distributions over a fixed number of topics and each topic has a multinomial distribution over words; Dirichlet priors make inference tractable but leave the topic number as a user-specified parameter. The HDP [75] extends this by placing Dirichlet process priors over topics, allowing the data itself to determine how many topics are needed—but at the cost of heavy computation, which has motivated parallel-inference schemes that unfortunately reintroduce fixed topic numbers per thread.

A second major limitation of both LDA and HDP is their bag-of-words assumption: Words are treated independently, so meaningful multi-word phrases (e.g., "machine learning") are split and lose their joint topical signal. Phrase-based topic models address this by first mining candidate phrases and then modeling their topic assignments. Early approaches forced every word in a mined phrase to share the same topic, but more recent work shows this is too restrictive—and yet the more flexible methods require complex, time-consuming phrase-mining steps.

In light of these considerations, we aim to develop a new parallel Gibbs sampling algorithm for nonparametric and coherent topic discovery.

2.1.3.2 Methodology

Precondition for Parallelization The existing parallel G2PP [16] aims to update model parameters independently via multi-threading by associating each topic with one thread. For $\boldsymbol{x}_d$, the number of words assigned to the k-th topic $n_{d,k}$ in G2PP must satisfy $\sum_{k=1}^{K} n_{d,k} = N_d$, which does harm to parallelization. For breaking the dependency between $n_{d,k}$ and N_d, a "copy" of $\boldsymbol{x}_d$ denoted as $\boldsymbol{x}'_d$ is used to "not observe" N_d in the parallel G2PP. $\boldsymbol{x}'_d$ is stored in a word stack S_d, and each element in $\boldsymbol{x}'_d$ is randomly drawn from $\boldsymbol{x}_d$ with replacement. Such a bootstrap technique that performs subsampling on the original document is a major precondition for parallelization, rendering infrequent words that help capture the full context of a topic unlikely to be sampled.

Adjustment of Topic Numbers Although using a fixed topic number in the parallel G2PP eases the management of threads, it cannot deal with the following two cases. First, G2PP adopts a concentration parameter to control the probability of adding a topic, corresponding to all the current topics reject to accept certain words in the word stack. For the parallel model, there should also be a chance to generate a new topic. Second, some marginal topics may accept few words, and the associated threads are seldom updated since the prior weights of topics are imbalanced in G2PP. Such topics are either redundant or unimportant, and they should be deleted to reduce the computation and memory costs. In our TSAM, we introduce two counting variables (*add_cnt* and *del_cnt*) to determine the new topic number. For each word stack, if its current size is larger than 0, then add_cnt++. For the kth topic, if both the current number of assigned words $n_{d,k}^{(0)}$ and the number of assigned words after ϵ epochs $n_{d,k}^{(\epsilon)}$ are less than a given proportion of the total number of words in the corpus, then del_cnt++. Based on an initial topic number K, we can simply adjust K by $add\_cnt - del\_cnt$. Note that reducing K corresponds to destructing the associated threads, and a monitor-executor mechanism is further proposed to avoid memory leak using a two-level hierarchical structure of threads. For all topics in TSAM, the assigned words can be asynchronously updated by a parallel Gibbs sampling algorithm based on the reversible jump Monte Carlo Markov Chain [25].

Copula Guided Topical Word Assignments To enhance the coherence of topics, we also equip our TSAM with a copula guided parallel Gibbs sampling algorithm and name the extended model as TSAMcop. The overall framework of our TSAMcop is shown in Fig. 2.6, and the data flows of adding and removing words in TSAMcop are illustrated in Fig. 2.7, where each X'_{di} and B'_{di} is maintained by one topic/thread t_i, $i \in \{1, \ldots, K\}$ for the dth subset. As aforementioned, the words that compose a phrase are grammatically related, and they should be highly likely to be assigned to the same topic. Inspired by the autoregressive models, we assume the topical dependencies within each phrase to be nested. Take a phrase with L words

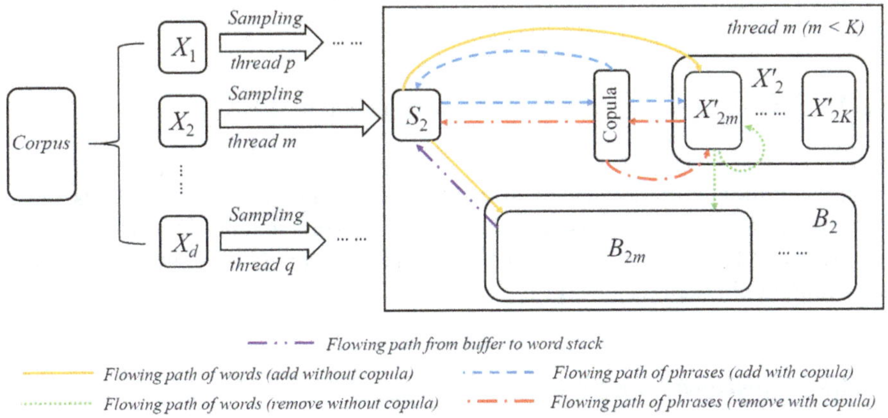

Fig. 2.6 The overall framework of TSAMcop, where the inference procedure inside thread m ($m < K$) is illustrated, and it is identical for other threads

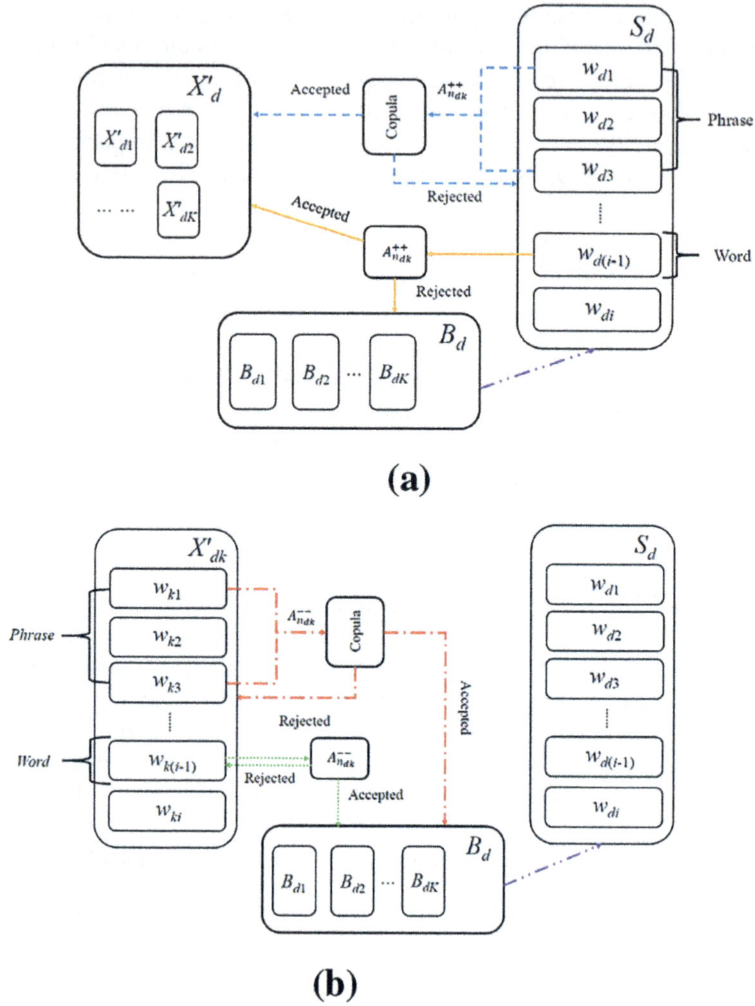

Fig. 2.7 The data flows of (**a**) adding and (**b**) removing words in TSAMcop

as an illustration, a Markov chain of order 1 is formed by the first two words, and a Markov chain of order $L-1$ is obtained by the last word conditioned on all previous words. In our TSAMcop, we construct a nested Frank copula for topical dependency modeling within phrases.

2.1.3.3 Experimental Results

In quantitative experiments, we use perplexity and topic coherence (measured by NPMI) to evaluate different methods. The results show that TSAMcop achieves better perplexity and NPMI scores than TSAM and other baseline models. This validates the effectiveness of copula guided models in boosting the generative probabilities of these corpora and the coherence of discovered topics. It is also interesting that both TSAM and TSAMcop obtain similar numbers of topics with HDP on three real-world corpora, which illustrates that our proposed method of adjusting topic number holds similar nonparametric characteristic with HDP.

To demonstrate the superiority of our proposed models in terms of convergence speed, we run our models and other baseline methods on NYTimes dataset. The results show that our models can converge as fast as the parallel G2PP on a large-scale corpus, while HDP and CopulaLDA failed to converge in 48 hours.

2.1.4 Label-Conditioned Neural Topic Model

Neural topic modeling is a scalable automated technique for text data mining. In its downstream tasks like content analysis, there is a focus on how well the discovered topics align with labels. However, due to the lack of guidance from labels, unsupervised neural topic models are less powerful in this regard. Existing supervised neural topic models often assume that labels are generated from topics and use document-topic distributions for a classifier to achieve label-topic alignment indirectly. Such a mechanism faces the following issues: (1) The assumption that labels are generated from topics leads to topics blending the latent patterns of multiple labels, which limits the label-topic alignment, and (2) the aforementioned alignment is derived from the classification task; thus the quality relies heavily on the performance of classifiers. To tackle these problems, we develop a novel supervised neural topic model by aligning topics with labels directly. It utilizes a chain-structured graphical model that emphasizes direct constraints of labels on topics and introduces soft indicators to explicitly construct the label-topic relationships. Besides, it can be seamlessly integrated with most existing unsupervised neural topic models in a flexible manner. Experiment results on multiple datasets demonstrate that our model can largely enhance the alignment between labels and topics while maintaining good topic quality.

2.1.4.1 Introduction

Topic modeling [5, 73] aims to uncover the latent semantic structure within a corpus, which has been applied in various downstream tasks, such as social event analysis [89] and opinion mining [77]. A topic is typically described as a probability distribution over words. When conducting topic discovery, we often focus on the extent to which the discovered topics are meaningful and easy to understand, i.e., the interpretability of topics. To assess it, the topic coherence and many other evaluation have been developed [43, 69]. However, a high coherence score of topics is often not sufficient enough to indicate that they are "good" topics [30, 31]. Considering that meaningful labels describe some inherent patterns within documents, we expect that the implicit codes of documents (i.e., topics) obtained by the topic model are aligned with their explicit codes (i.e., labels). Achieving label-topic alignment facilitates better understanding of the discovered topics, especially in the context of content analysis [31]. Unfortunately, due to the lack of guidance from manually defined labels, it is challenging for topics discovered by unsupervised topic models to be consistent with them. As a result, these topics fail to effectively capture underlying patterns shared among documents with the same label. Additionally, it becomes challenging for human, especially for interested users lacking relevant domain expertise, to interpret misaligned topics in the context of human-defined labels [30]. In response to these issues, several supervised topic models have been developed [3, 12]. The fundamental idea behind most of them is to predict labels through document representations (i.e., document-topic distributions), forcing topic distributions to distinguish between different document categories, thereby indirectly aligning topics with labels. Nevertheless, they are confronted with the following challenges: (1) The assumption that labels are generated from topics leads to topics blending the latent patterns of multiple labels. As training progresses, topics capture latent semantics from documents of multiple categories, for which the label-topic alignment is limited. (2) For these models, label-topic alignment relies heavily on the classification performance. Since it is challenging to conduct text classification on imbalanced datasets [58], existing supervised topic models may have difficulty mining topics that align with minority categories.

While there are statistical topic models based on Gibbs sampling that explicitly construct the relationship between topics and labels [67], they suffer from inflexible model-specific derivations and slow inference speed. They are not neural topic models (NTMs) and do not leverage the advantages of deep generative models like Variational Auto-Encoder (VAE) [41, 73]. As summarized in [31], NTMs can benefit from pretraining on extensive and diverse data; thus they are straightforward to use multilingually or multimodally. NTMs can also be integrated for joint modeling within larger neural architectures smoothly.

To achieve label alignment on the premise of retaining the advantages of NTMs, we propose a novel supervised NTM, named Label-Conditioned Neural Topic Model (LCNTM), by explicitly constructing the label-topic relationships. First, within the framework of VAE, we introduce a novel chain-structured graphical model that emphasizes direct constraints of labels on topics. This differs from the assumption in most supervised topic models, such as SCHOLAR [12], where labels are generated from topics. Through derivation, we obtain the corresponding evidence lower bound (ELBO), which involves variational distribution and prior conditioned on labels. Second, we introduce soft indicators to assign labels to topics and thus explicitly model the label-topic relationships. We then design normal variational distribution and prior that align with this mechanism, thereby masking out topics unrelated to the labels. In addition, our proposed model can be seamlessly integrated with most existing unsupervised NTMs. Experiment results on multiple datasets demonstrate that our model can effectively enhance the alignment between labels and topics while maintaining topic quality. The main contributions of this study are as follows:

- We propose a novel graphical model for supervised neural topic modeling. In the generative process, we emphasize the control of labels over topics, aiming to achieve stronger alignment between labels and topics.
- We introduce soft indicators to model the variational distribution and prior conditioned on labels and explicitly construct the label-topic relationships.
- The proposed model can be easily integrated with most existing unsupervised NTMs.
- Experiment results on multiple datasets show that our proposed model can achieve better label-topic alignment without compromising the topic quality.

In Sect. 3.2.1, we will enhance LCNTM by adopting more powerful techniques like optimal transport to learn label-topic relationships adaptively.

2.1.4.2 Methodology

In this section, we propose LCNTM, a novel supervised NTM. We begin by describing the background and the generative process. Then, in order to model the variational distribution and prior that conditioned on labels, we introduce soft indicators, explicitly constructing the label-topic relationships. Ultimately, we describe the parameter inference for LCNTM and discuss how to conduct inference in the absence of labels.

Background Consider a document collection $\mathcal{X}$ with V unique words. Each document in $\mathcal{X}$ is typically represented by a bag-of-words (BoW) vector $\boldsymbol{x} \in \mathbb{R}^V$ and a latent distribution over K topics (document-topic distribution), denoted by $\boldsymbol{z} = (z_1, \ldots, z_K) \in \mathbb{R}^K$. Each topic describes a specific semantic concept, and the kth topic is regarded as a distribution over words (topic-word distribution), denoted by $\boldsymbol{\beta}_k \in \mathbb{R}^V$. $\boldsymbol{\beta}_k$ is often obtained through topic embeddings $\mathbf{T} \in \mathbb{R}^{H\times K}$ and word embeddings $\mathbf{W} \in \mathbb{R}^{H\times V}$ in an H-dimensional embedding space, which is a common practice in state-of-the-art NTMs [18, 86, 94].

NTMs utilize neural networks to infer the latent document-topic distribution [73]. Most existing NTMs are based on the VAE framework, where latent variables are interpreted as topics. An unsupervised VAE-based NTM typically assumes that the latent document-topic distribution $\boldsymbol{z}$ lying within a probability simplex $\Delta^K = \left\{\boldsymbol{z} \in \mathbb{R}^K_{\geq 0} | \sum_k z_k = 1\right\}$ is drawn from a prior distribution $p(\boldsymbol{z})$; then the BoW vector $\boldsymbol{x}$ is generated conditioned on $\boldsymbol{z}$, i.e., from $p(\boldsymbol{x}|\boldsymbol{z})$. In this setting, the joint distribution $p(\boldsymbol{x}, \boldsymbol{z})$ can be factorized by $p(\boldsymbol{x}, \boldsymbol{z}) = p(\boldsymbol{z})p(\boldsymbol{x}|\boldsymbol{z})$. To approximate the intractable posterior distribution and thus calculate ELBO, the variational distribution $q(\boldsymbol{z}|\boldsymbol{x})$ is introduced in VAE.

Supervised NTMs involve the label of each training document in the generative process. Suppose that the label set $\mathcal{Y}$ has C labels. Consider the observations $\mathcal{O} = \{(\boldsymbol{x}_i, y_i)\}_{i=1}^N$ in the context of supervised NTM, where $y \in \mathcal{Y}$ denotes the singly assigned label of $\boldsymbol{x}$. SCHOLAR [12], a classic supervised NTM utilizing a forked graphical model $y \leftarrow z \rightarrow \boldsymbol{x}$, assumes that labels are generated from topics. This implies that z is drawn from $p(z)$ without being constrained by y in the generative process.

It is questionable whether such z in SCHOLAR can actually capture topic information aligned with y. This motivates us to rethink the generative process and finally propose a novel supervised NTM, where z is drawn from $p(z|y)$.

Generative Process Different from the forked graphical model adopted by SCHOLAR, we instead choose the graphical model of LCNTM to be a *chain*, i.e., $y \rightarrow z \rightarrow \boldsymbol{x}$. Thus, the joint distribution $p(\boldsymbol{x}, y, z)$ can be factorized as follows:

$$p(\boldsymbol{x}, y, z) = p(y)p(z|y)p(\boldsymbol{x}|z). \tag{2.8}$$

The intuition behind the chain-structured graphical model is that **the topics are generated under the constraint of the labels**. Through the strong constraint imposed by the labels, the generation of topics (modeled by $p(z|y)$) becomes specific to each label, thereby explicitly making the topics distinctive in terms of document labels. Figure 2.8 shows the graphical models of multiple NTMs, including the classic unsupervised NTM (i.e., ProdLDA) [73], the supervised NTM (i.e., SCHOLAR) [12], and our proposed LCNTM.

Evidence Lower Bound For $p(y)$, we define it to be discrete uniform distribution, i.e.,

$$p(y) = \text{Cat}(\mathbf{1}_C/C), \tag{2.9}$$

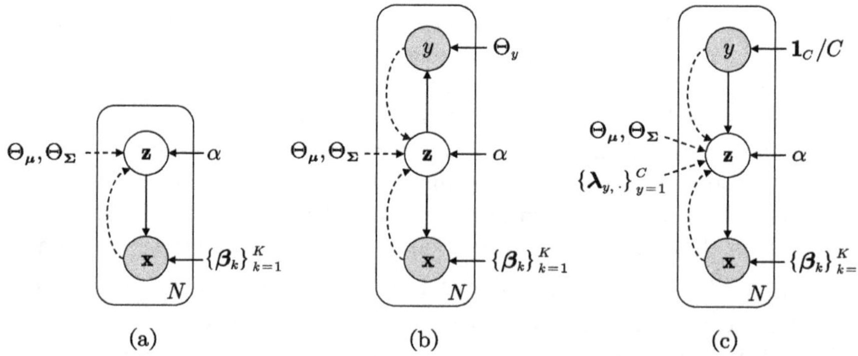

Fig. 2.8 Graphical models of different topic models. Solid arrows denote generative process, and dashed arrows denote inference for all models. (**a**) ProdLDA, (**b**) SCHOLAR, and (**c**) LCNTM (ours)

where $\mathrm{Cat}(\cdot)$ denotes the categorical distribution. $\mathbf{1}_C$ is a C-dimensional vector where all elements are set to 1. With this definition, we can derive the ELBO:

$$\begin{aligned}\log p(\boldsymbol{x}, y) &= \log\left(\mathbb{E}_{q(z|\boldsymbol{x},y)}\left[\frac{p(\boldsymbol{x}, y, z)}{q(z|\boldsymbol{x}, y)}\right]\right)\\ &\geq \mathbb{E}_{q(z|\boldsymbol{x},y)}\left[\log p(\boldsymbol{x}, y, z) - \log q(z|\boldsymbol{x}, y)\right]\\ &= \mathbb{E}_{q(z|\boldsymbol{x},y)}\left[\log p(\boldsymbol{x}|z) + \log p(z|y) + \log p(y) - \log q(z|\boldsymbol{x}, y)\right]\\ &= \mathbb{E}_{q(z|\boldsymbol{x},y)}\left[\log p(\boldsymbol{x}|z)\right] - D_{\mathrm{KL}}\left[q(z|\boldsymbol{x}, y) \parallel p(z|y)\right] + \log p(y).\end{aligned} \tag{2.10}$$

Note that $\log p(y)$ is a constant because we choose $p(y)$ to be a discrete uniform distribution, as described in Eq. (2.9).

To maximize the ELBO, we use the following loss function:

$$\mathcal{L}_{\mathrm{VAE}}(\boldsymbol{x}, y) = D_{\mathrm{KL}}[q(z|\boldsymbol{x}, y) \parallel p(z|y)] - \mathbb{E}_{q(z|\boldsymbol{x},y)}[\log p(\boldsymbol{x}|z)], \tag{2.11}$$

where D_{KL} denotes the KL divergence between two distributions. $\mathcal{L}_{\mathrm{VAE}}$ is similar in form to the loss function of the original VAE [41], as both are composed of a KL divergence term and a reconstruction term, and document reconstruction relies solely on latent variables z (i.e., topics). The difference between the two lies in that our supervised model utilizes a variational distribution and a prior additionally conditioned on label y.

Aligning Topics with Labels How can we model $p(z|y)$ to achieve better label-topic alignment? A natural choice is to have each topic contain only the latent semantics of documents under a certain category. In other words, we hope the document representations inferred from documents of different categories to be *orthogonal*. To this end, we introduce soft indicators and utilize them to model $q(z|\boldsymbol{x}, y)$ and $p(z|y)$.

Inspired by Labeled LDA [67], we use a **soft indicator** $\lambda_{y,k} > 0$ for the kth topic and label y to explicitly build the label-topic relationships.[2] A higher value of $\lambda_{y,k}$ indicates a higher degree of belongingness of the kth topic to label y. A fundamental idea is that when the kth topic is not under label y, then the kth dimension of the inferred latent topic distribution z_k should be masked by setting it to 0; conversely, it should be retained. To integrate this idea into the NTM with logistic normal prior, we model $q(z|\boldsymbol{x}, y)$ and $p(z|y)$ by

$$q(z|\boldsymbol{x}, y) = \mathcal{LN}(\boldsymbol{\mu} + \ln \boldsymbol{\lambda}_{y,\cdot}, \boldsymbol{\Sigma}), \tag{2.12}$$

$$p(z|y) = \mathcal{LN}(\boldsymbol{\mu}_0 + \ln \boldsymbol{\lambda}_{y,\cdot}, \boldsymbol{\Sigma}_0), \tag{2.13}$$

[2] With a slight abuse of notation, we use y to denote the corresponding label's index.

where $\mathcal{LN}(\cdot, \cdot)$ denotes the logistic normal distribution. $\boldsymbol{\lambda}_{y,\cdot} = (\lambda_{y,1}, \ldots, \lambda_{y,K})$ is the indicator vector for y with respect to each topic. $\boldsymbol{\mu} = f_{\mu}(\boldsymbol{x}; \Theta_{\mu})$ and $\boldsymbol{\Sigma} = f_{\Sigma}(\boldsymbol{x}; \Theta_{\Sigma})$ are the mean and diagonal covariance matrix encoded by $\boldsymbol{x}$, while $\boldsymbol{\mu}_0$ and $\boldsymbol{\Sigma}_0$ are determined by Laplace approximation [27] with hyperparameter α. To see how it works, consider the latent normal logits $\boldsymbol{r} = (r_1, \ldots, r_K) \sim \mathcal{N}(\boldsymbol{\mu}, \boldsymbol{\Sigma})$ that are independent of y. We draw the document-topic distribution $\boldsymbol{z}$ from $q(\boldsymbol{z}|\boldsymbol{x}, y)$ with $\boldsymbol{r}$ by

$$z_k = \left[\text{Softmax}(\boldsymbol{r} + \ln \boldsymbol{\lambda}_{y,\cdot})\right]_k = \frac{\lambda_{y,k} \exp(r_k)}{\sum_{k'=1}^{K} \lambda_{y,k'} \exp(r_{k'})}. \tag{2.14}$$

We can see that, after the softmax transformation, the additive term $\ln \lambda_{y,k}$ becomes the coefficient for $\exp(r_k)$. This coefficient dominates the allocation of each topic in $\boldsymbol{z}$, with topics not under y being assigned almost negligible weights.

Label Assignment for Topics We can determine the label-topic relationships in a flexible manner. When we focus on ensuring that the kth topic captures the semantics of only one particular label y, it is natural to set $\lambda_{y,k}$ to 1 and $\lambda_{y,k'}$ to a small positive real number like 10^{-10} for all $y' \neq y$ so that the kth topic is exposed only to documents with label y. In this context, we refer to this practice as *assigning labels to topics*. This is also our motivation, namely to achieve better label-topic alignment through strong constraints imposed by labels on topics. For convenience, we denote the number of topics under label y as η_y.

But the potential of soft indicators goes beyond this. If we are interested in semantic patterns shared by different labels, namely, common topics, we can assign multiple labels to one topic. For example, if we want to examine the common topics between two similar categories y_1 and y_2, we can set both $\lambda_{y_1,k}$ and $\lambda_{y_2,k}$ to 1. As an extension of the proposed method, we demonstrate this scenario in the experiments.

In addition, the mechanism of soft indicators makes it possible to dynamically learn the label-topic relationships. We leave this in Sect. 3.2.1.

Applicable to Various Decoders As mentioned above, our proposed model concentrates on inference rather than decoding, which involves reconstructing documents using the inferred document-topic distributions and topic-word distributions. This implies that our model can be combined with decoders of various existing NTMs, such as ETM [18] and ECRTM [86]. In our experiments, we construct our LCNTM based on ETM and ECRTM, respectively. In order to achieve maximum label-topic alignment, we let each topic belong to one label, and the number of topics under one certain label follows the label distribution of the dataset.

Inference in the Absence of Labels One should feed $\boldsymbol{x}$ and y simultaneously to LCNTM to infer the document-topic distribution $\boldsymbol{z}$. But at the testing phase, y is absent. To deal with it, we follow [12] and use the label $\hat{y}$ that maximizes the likelihood as an estimate for ground truth label y after LCNTM is trained, i.e.,

$$\hat{y} = \underset{y' \in \mathcal{Y}}{\operatorname{argmax}} \log p(\boldsymbol{x}, y') \approx \underset{y' \in \mathcal{Y}}{\operatorname{argmin}} \mathcal{L}_{\text{VAE}}(\boldsymbol{x}, y'). \tag{2.15}$$

Then we can infer the document-topic distribution of $\boldsymbol{x}$ using $\hat{y}$.

2.1.4.3 Experimental Results

We conduct experiments on the following benchmark datasets: 20NewsGroup (20NG), IMDB, Yahoo! Answer (Yahoo), and Seven Categories (SC). 20NG and IMDB are relatively long text datasets, with 20 and 2 labels, respectively. Yahoo is a short text dataset with ten labels, while SC is imbalanced over seven categories, with the proportion of documents in different categories ranging from 3.1% to 24.6%. To be consistent with the datasets used in [12, 86], we adopt the same preprocessing approach and set the vocabulary size to 5000. As SC does not come with predefined training set and test set, we conduct stratified random sampling and sample 20% of the documents from the entire collection to serve as the test set. The statistics of these datasets in the experiments are provided in Table 2.3.

We compare our LCNTM with the classic supervised NTM, **SCHOLAR** [12], and two unsupervised state-of-the-art NTMs: **ETM** [18] and **ECRTM** [86]. SCHOLAR learns label-relevant topics by using document-topic distributions to predict ground truth labels. ETM is a classic NTM that models words and topics in the same embedding space. ECRTM is a state-of-the-art NTM that avoids topic collapsing by embedding clustering regularization.

Label-Topic Alignment As discussed in the previous study [31], inter-coder reliability is one of the central considerations in content analysis. Therefore, we evaluate the alignment between labels (explicit codes of documents assigned by human) and topics (implicit codes given by topic models) across different models. A common practice [31, 86, 94] is to cluster the document-topic distributions to determine the categories induced by the topic model and assess the extent to which these categories align with the ground truth labels using clustering metrics. Here, we calculate the following three cluster quality metrics: Adjusted Rand Index (ARI), Purity, and Normalized Mutual Information (NMI). We use two clustering strategies on the document-topic distributions of testing documents: (1) apply **k-means** algorithm and (2) use the most dominant topic as the assigned cluster (referred to as **top** strategy). Similar to [82], we set the number of clusters to

Table 2.3 Statistics of the datasets after preprocessing

Dataset	#Docs (Train)	#Docs (Test)	Vocab. size	Avg. length	#Labels
20NG	11,314	7532	5000	110.5	20
IMDB	25,000	25,000	5000	95.0	2
Yahoo	10,000	2500	5000	35.4	10
SC	2505	622	5000	170.6	7

Table 2.4 ARI, Purity, and NMI scores under 50 and 100 topics

Dataset	Model	K = 50						K = 100					
		kmeans			top			kmeans			top		
		ARI	Purity	NMI	ARI	Purity	NMI	ARI	Purity	NMI	ARI	Purity	NMI
20NG	SCHOLAR	0.097	0.358	0.424	0.320	0.591	0.527	0.071	0.265	0.361	0.242	0.629	0.510
	ETM	0.022	0.293	0.346	0.260	0.575	0.471	0.011	0.237	0.281	0.194	0.579	0.457
	LCNTM+ETM	0.522	0.714	0.627	0.367	0.722	0.577	0.502	0.687	0.611	0.235	0.734	0.548
	ECRTM	0.113	0.408	0.463	0.374	0.574	0.521	0.049	0.317	0.377	0.322	0.559	0.487
	LCNTM+ECRTM	**0.577**	**0.751**	**0.665**	**0.429**	**0.761**	**0.622**	**0.581**	**0.752**	**0.667**	**0.360**	**0.762**	**0.602**
IMDB	SCHOLAR	0.063	0.675	0.091	0.014	0.761	0.085	0.009	0.572	0.076	0.008	0.767	0.076
	ETM	0.006	0.603	0.050	0.010	0.653	0.025	0.004	0.543	0.007	0.000	0.501	0.001
	LCNTM+ETM	**0.079**	0.849	0.160	0.029	0.847	0.128	**0.065**	**0.793**	**0.125**	0.014	**0.794**	0.082
	ECRTM	0.021	0.666	0.061	0.012	0.686	0.055	0.025	0.656	0.058	0.005	0.695	0.046
	LCNTM+ECRTM	0.055	**0.860**	**0.162**	**0.042**	**0.860**	**0.145**	0.039	0.784	0.092	**0.031**	0.788	**0.093**
Yahoo	SCHOLAR	0.009	0.253	0.197	0.113	0.542	0.305	0.006	0.222	0.162	0.081	0.569	0.312
	ETM	0.013	0.280	0.203	0.092	0.521	0.272	0.005	0.229	0.158	0.053	0.535	0.276
	LCNTM+ETM	**0.263**	0.621	0.370	0.131	0.625	0.340	**0.297**	0.617	0.366	0.073	0.622	0.328
	ECRTM	0.101	0.459	0.317	0.121	0.545	0.294	0.033	0.319	0.248	0.103	0.561	0.304
	LCNTM+ECRTM	0.262	**0.633**	**0.378**	**0.134**	**0.635**	**0.352**	0.253	**0.618**	**0.367**	**0.109**	**0.624**	**0.339**
SC	SCHOLAR	0.070	0.613	0.455	0.259	0.961	0.641	0.033	0.512	0.373	0.195	0.959	0.583
	ETM	0.084	0.612	0.448	0.193	0.969	0.617	0.027	0.481	0.336	0.120	0.974	0.565
	LCNTM+ETM	**0.507**	0.953	0.732	0.191	0.989	0.628	**0.518**	0.932	0.725	0.120	0.987	0.576
	ECRTM	0.333	0.878	0.629	**0.265**	0.940	0.641	0.112	0.627	0.443	0.196	0.966	0.621
	LCNTM+ECRTM	0.449	**0.989**	**0.748**	0.262	**0.990**	**0.672**	0.420	**0.991**	**0.741**	**0.210**	**0.990**	**0.643**

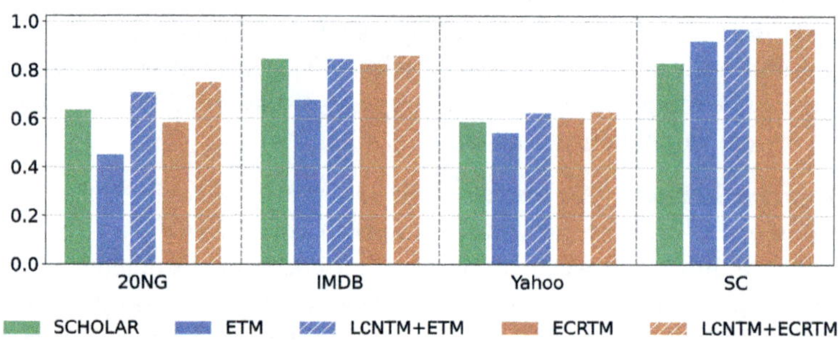

Fig. 2.9 Macro F1 scores in testing the quality of label-topic alignment. We run all models five times and report the average results

20 in k-means algorithm. We also report the macro F1 scores of each model on the classification task to further explore the alignment between labels and topics. For LCNTM, we directly predict labels by Eq. (2.15). For SCHOLAR and other unsupervised baselines, we use the document-topic distributions as features for documents and train SVM classifiers. The results on the clustering and classification metrics are presented in Table 2.4 and Fig. 2.9, respectively. For clarity, the best scores across all models are in **bold**. The blue cell color in the rows corresponding to LCNTM indicates the score superior to that of the corresponding unsupervised model. We run all models five times and report the average results. Unless otherwise mentioned, we do not cite the results reported in ECRTM [86] but rerun all the baselines, so as to obtain ARI scores and all results on the label-imbalanced SC dataset.

From the experiment results, we can draw the following conclusions: (1) **Our proposed model can effectively improve the label-topic alignment of unsupervised NTMs**. Despite the less significant improvements (or even a slight decrease) of LCNTM on a few metrics on SC, we believe this might be due to the mismatch between the optimal number of topics within each category and η we set for SC (i.e., the label distribution). Due to the lack of label guidance, unsupervised NTMs cannot autonomously achieve alignment between labels and topics. On the contrary, during training, our proposed models mask irrelevant topics with respect to the labels assigned by human and thereby explicitly force each topic to capture underlying patterns shared among documents with the same label, resulting in achieving higher label-topic alignment. (2) **Compared to the existing supervised NTM, SCHOLAR, our proposed model is able to learn topics that are more aligned with labels**. Thanks to the direct construction of the label-topic relationships in our proposed model and the powerful ability of ECRTM to mitigate topic collapsing, LCNTM+ECRTM consistently outperforms SCHOLAR on all datasets. In addition, we can also see that LCNTM+ETM outperforms SCHOLAR on most datasets. (3) **The topics learned by LCNTM are capable of effectively distinguishing between different labels, as evidenced in the classification experiment**. Based solely on the maximum evidence lower bound obtained under different label assumptions to determine the category, it is sufficient to achieve better classification performance than using the document-topic distribution as the input of a popular classifier (i.e., SVM). This illustrates that the topics learned under different labels succeed in capturing the shared patterns among documents associated with different labels.

Topic Quality To evaluate the topic quality, we consider both Topic Coherence (TC) and Topic Diversity (TD). TC measures the coherence of the top words within a topic. We consider the top-10 words for each topic and calculate C_V [69] with Wikipedia article collection as the reference corpus. Following [18], we calculate the proportion of unique words among the top-25 words for each topic as TD. Figure 2.10 showcases the topic quality of the evaluated models, where all models are run five times and the averaged results are reported.

The results show that **the topic quality is maintained after applying our proposed model**. For ETM, both TC and TD show improvement after applying LCNTM on 20NG, Yahoo, and IMDB datasets, while on SC there is a trade-off between TC and TD. For ECRTM, such trade-off is more prevalent across different datasets. Indeed, the alignment between labels and topics does not necessarily imply an improvement in topic quality. The soft indicators selectively expose topics to documents of different categories and thus achieve the alignment between labels and topics. But this is neither directly related to the extraction of word co-occurrence patterns (TC) nor to the issue of topic collapsing (TD).

Topic Discovery on Imbalanced Datasets In this part, we concentrate on imbalanced dataset SC and examine the top-10 topic words discovered by SCHOLAR, LCNTM+ETM, and LCNTM+ECRTM to showcase the benefits

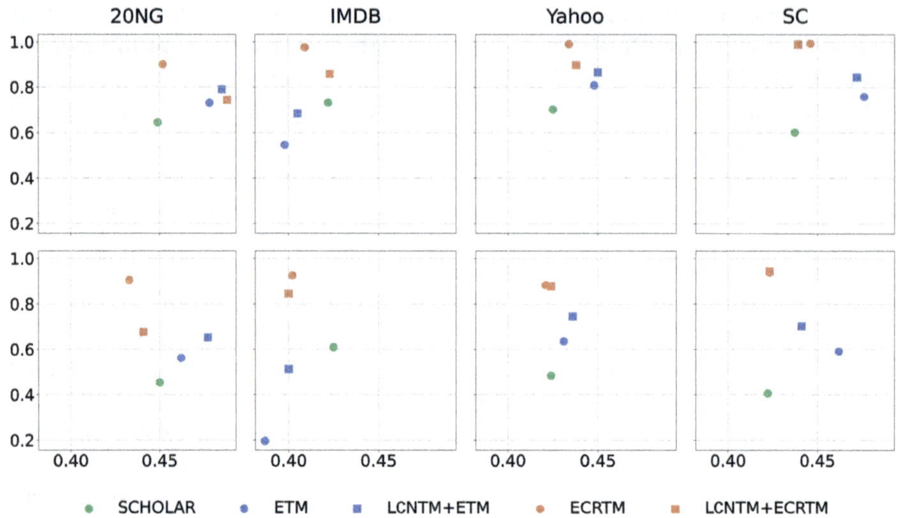

Fig. 2.10 Topic quality of evaluated models showing TC (C_V) versus TD relationships with $K = 50$ (top row) and $K = 100$ (bottom row)

brought by explicit label-topic relationships. Specifically, we explore three of the rarest labels: `geography` (3.1%), `maths` (6.9%), and `accounts` (9.1%). Other categories include `history` (15.6%), `biology` (20.3%), `computer_science` (20.4%), and `physics` (24.6%). Although SCHOLAR does not explicitly specify the label-topic relationships, we can still feed a one-hot vector, which has 1 for the kth dimension and 0 for others, to the classifier and determine which label the kth topic belongs to. We show the mined topics for different models under the rare categories in Table 2.5, where each row represents a topic. When we run SCHOLAR, we find that in most runs, SCHOLAR *fails* to mine topics under `geography`. Therefore, we run SCHOLAR for multiple times to report the topics under `geography`. As shown in Table 2.5, the highlighted topics in bold, respectively, exhibit potential entanglement with `biology`, `physics`, and `maths`, while topics discovered by LCNTM+ETM and LCNTM+ECRTM show the uniqueness of the rare categories. We believe that due to the limitations imposed by performance of classifiers, SCHOLAR struggles to focus on minority categories and extract their unique topics; the few topics under minority categories also tend to be entangled with those under majority categories. In contrast, our proposed LCNTM concentrates on the encoding of documents from minority classes into specific topics, effectively disentangling these topics from those under other labels.

Discovering Common Topics Under Multiple Labels As discussed above, the mechanism of soft indicators makes it possible to discover common topics under multiple labels. To see this, we run LCNTM+ETM on 20NG. In particular, we focus on four labels: `rec.autos`, `rec.motorcycles`, `rec.sport.baseball`, and `rec.sport.hockey`. They are all related to recreation. For these labels, we assign two topics. Besides, we set two common topics. For these common topics and

Table 2.5 Top-10 topic words for topics under minority categories in SC

Geography	
SCHOLAR	Forests forest rivers biodiversity temperate tropical lakes geography toxic deserts
LCNTM+ ETM	Geography settlements biodiversity species ecosystem villages urban roads houses fundamentals
	Moisture route warm ocean rainfall wind clouds vapour currents humidity
LCNTM+ ECRTM	Urban towns Panama continent employment pacific houses educational infrastructure Paris
	Oceans circulation relief tropical precipitation cold hemisphere violent humidity evaporation
Maths	
SCHOLAR	Digit algorithm remainder digits decimal numbers ten fraction integers divisor
	Denominator resistors voltage biology molecular semiconductor lesson resistor circuit heat
	Algebra solve homework helpers math lesson sides triangle inches simplify
LCNTM+ ETM	Variable inequalities coordinate ordered slope algebraic ray graphs points graph
	Simplify solve expressions sides exponents inequality mark price equation multiply
	Fractions denominator decimal fraction numerator integers numbers evenly negative lesson
	Triangle inches leg lesson length legs perimeter review lengths algebra
LCNTM+ ECRTM	Proportions percent simplify renewed expressions Africans intensified centers Corfield funding
	Perimeter inches algebra triangle inequalities triangles geometry ordered circle lengths
	Triangle perimeter triangles inches ordered inequalities algebra geometry strict rep
	Decimal decimals prime fractions numbers evenly subtracting integers easiest numerator
Accounts	
SCHOLAR	Cash purchases sales credit ledger debit recording transactions depreciation recorded
	Review lesson algebra homework solution math helpers exponents simplify biology
	Allotment shares premium share debentures money issued equity debenture application
LCNTM+ ETM	Cash cheque particulars date transactions bank sales goods paid traders
	Sales account credit posted suspense debit errors trial posting purchases
	Shares share allotment equity money application capital premium company call
	Cash debentures assets liabilities redemption activities equivalents tax investing debenture
LCNTM+ ECRTM	Cash financing income depreciation expenses loans receivables equivalents payables tax
	Discount ltd redemption particulars debenture debentures secured date loan issued
	Ledger credit account suspense debit debited stationery furniture purchases transactions
	Shares capital allotted refunded rata subscription reissued equity applicants money

Table 2.6 Top-10 topic words for topics under labels related to *recreation*

common	bos det tor pit que cal min calgary montreal mon
	baseball chris rocks posting thanks bmw wondering jay rec mike
rec.autos	car saturn cars toyota dealer honda engine dodge ford oil
	cars brake car tires radar automotive mustang brakes tire wheel
rec.motorcycles	bike bikes dod bmw motorcycle honda riders exhaust ama motorcycles
	helmet bike dog ride dod riding motorcycle levine drinking tire
rec.sport.baseball	cubs phillies sox yankees mets braves suck philadelphia won games
	hitter pitching players morris baseball pitchers pitcher braves fame ball
rec.sport.hockey	islanders devils calgary captain pittsburgh rangers traded penguins winnipeg abc
	puck players flyers nhl hockey sharks team season teams player

the labels mentioned above, we set $\lambda_{y,k}$ to 1. Table 2.6 reports the top (i.e., representative) words for ten topics. We can observe that the common topics succeed in capturing the common aspects shared among these labels. For instance, in the first common topic, the top words are mainly about locations such as "bos" (Boston) and "det" (Detroit), which are not specific to any particular label. The second common topic is a mixture of aforementioned labels, containing words like "baseball," "BMW," as well as some names ("Chris," "Jay," and "Mike"). We can also see that the topics under each label describe distinctive aspects of the label. This demonstrates the flexibility of the soft indicator mechanism and its applicability across various scenarios besides achieving label-topic alignment.

2.1.5 Summary of Flat Topic Models

In summary, there are many problems remained to be solved in traditional flat topic models, such as reasonable priors, effective computation, and the utilization of metadata. In order to discover topics that align with labels, we propose SLTM and LCNTM. The former adopts label prediction and contrastive learning, while the latter adopts a label-conditioned prior to explicitly model label-topic relationships. To better match topic models and dispersed text datasets, we propose NMCMs, which uses the negative binomial distribution as the prior. In terms of effective computation, we propose TSAM and TSAMcop, both of which are nonparametric and adopt G2PP for efficient topic generation. Besides these, there are other promising methods remained to be searched in the future, like embedding modeling and introducing powerful pretrained/large language models.

2.2 Hierarchical Topic Models

Hierarchical topic models are essential tools in text mining for uncovering the latent semantic structures within corpora. By organizing topics into hierarchical structures, these models reveal the hierarchical relationships among topics, which is crucial for understanding the deep structure of document semantics. In this section, we first introduce our work entitled "Tree-Structured Topic Modeling with Nonparametric Neural Variational Inference" [16], "Nonparametric Forest-Structured Neural Topic Modeling" [93], and "Nonlinear Structural Equation Model Guided Gaussian Mixture Hierarchical Topic Modeling" [13]. Then, we present a preliminary study on generating a topic hierarchy with flexible topic trees, which can better discover the latent semantic structure of different corpora.

2.2.1 Tree-Structured Neural Topic Model

This part introduced a nonparametric Tree-Structured Neural Topic Model (nTSNTM) leveraging Neural Variational Inference (NVI). The model begins by learning the latent components of the **stick-breaking process** for each document and then models the affiliations of these latent components through **dependency matrices** between network layers. This network structure enables the efficient extraction of a tree-structured topic hierarchy that is reasonably structured, has low redundancy, and is adaptable in width.

2.2.1.1 Introduction

The tree-structured topic model [4] was devised to derive coherent topics from text while preserving their inherent hierarchical structure. This approach has demonstrated its utility in diverse downstream applications. For instance, it has been employed for hierarchical categorization of web pages [56], extraction of aspect hierarchies in reviews [39], and discovery of research topic hierarchies in academic repositories [64].

Despite their practical significance and potential benefits, tree-structured topic models confront several challenges. Firstly, the hierarchical arrangement of topics must be logical [79]. Generally, topics closer to the root are broader, while those nearer the leaves are more specialized. Additionally, child topics must align consistently with their parent topics. Secondly, it is essential to minimize redundancy in the extracted topics to prevent the distributions of parent and child topics from being overly similar [4]. Thirdly, the model should automatically determine the number of topics at each hierarchical level, as this is often unknown and cannot be predetermined [38]. Lastly, probabilistic models struggle with data scalability [35].

Although several tree-structured topic models have been developed [4, 35, 38], they have not fully addressed these challenges.

In this study, we concentrate on organizing topics into a reasonable tree structure using the Neural Variational Inference (NVI) framework [41, 68] with a nonparametric prior. Neural networks, known for their superior function fitting capabilities, have been increasingly integrated into topic modeling. However, few neural methods explicitly capture the dependencies among different layers and produce interpretable hierarchical topics, largely due to the limited interpretability of neural networks. Furthermore, the inflexibility of neural networks poses challenges in learning an unbounded number of topics at each level. To overcome these limitations, we introduce a novel nonparametric neural approach to generate tree-structured topic hierarchies, termed the nonparametric Tree-Structured Neural Topic Model (nTSNTM). By linking network layers with dependency matrices, our model can extract an interpretable tree-structured hierarchy. Specifically, the topic relationships across hierarchy levels are determined by the discrete vectors of the dependency matrices. To manage redundancy among topics, our model is allowed to freely generate topics without replicating their corresponding parent topics. Additionally, we integrate a stick-breaking process with NVI to enable the topic tree to have self-determined widths, facilitating automatic topic count determination. Finally, leveraging the strengths of neural networks, our model can easily scale to larger datasets. Experimental results demonstrate that our model outperforms baseline methods on several commonly used metrics and two newly developed measurements specifically for tree-structured topic models.

2.2.1.2 Methodology

In this section, we firstly describe the stick-breaking process. Then, we introduce the modeling of tree-structured topic hierarchy. Finally, we detail the inference method of our nTSNTM.

Stick-Breaking Process For nonparametric models, the stick-breaking prior is a random measure defined as $G = \sum_{k=1}^{\infty} \pi_k \delta_{\zeta_k}$, where δ_{ζ_k} represents a discrete measure concentrated at ζ_k, which is drawn from a base measure G_0 [34]. In the context of topic models, ζ_k can be interpreted as the kth topic, and G_0 represents the space of possible topics. The weights π_k are random and independent of G_0 [60]. This constructive definition is known as the Stick-Breaking Process (SBP) [70]. The SBP implies that the weights $\boldsymbol{\pi} = (\pi_k)_{k=1}^{\infty}$ can be generated by iteratively breaking off segments from a unit stick, where each segment represents the probability mass assigned to a particular topic. This process allows for a potential number of topics to be considered, with the weights determining the relative importance of each topic in the model.

As illustrated in Fig. 2.11, we fractionate the unit stick, resulting in the initial segment of length v_1. Subsequently, a portion v_2 of the residual stick is fractionated, yielding the second segment with length $v_2 \times (1 - v_1)$ and a leftover stick of

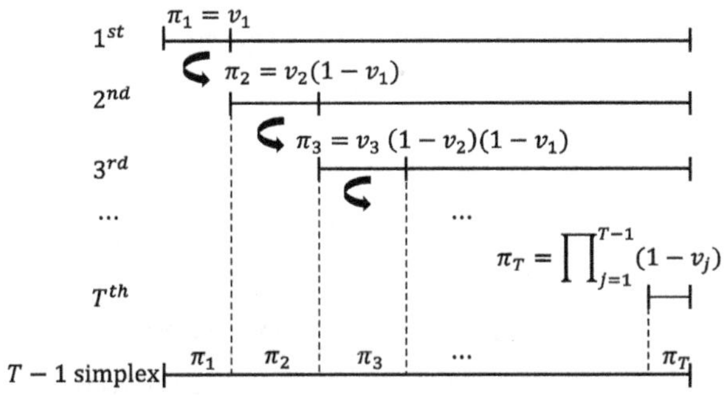

Fig. 2.11 Stick-breaking construction

length $(1 - v_1) \times (1 - v_2)$. This fractional breakdown continues iteratively on the remaining stick. Assuming a truncation threshold T, the ultimate segment's length will be given by $\prod_{j=1}^{T-1}(1 - v_j)$. Mathematically, the length of each segment can be formulated as

$$\pi_k = \begin{cases} v_1 & \text{when } k = 1, \\ v_k \cdot \prod_{t<k}(1 - v_t) & \text{for } k > 1. \end{cases} \tag{2.16}$$

In the above, each v_k is drawn from a Beta distribution parameterized by α_0 and β_0, where α_0 and β_0 represent the prior parameters. It is worth noting that the segment weights $\boldsymbol{\pi}$ adhere to the constraints $0 \leq \pi_k \leq 1$ and $\sum_{k=1}^{\infty} \pi_k = 1$, allowing us to interpret $\boldsymbol{\pi}$ as a set of random probabilities. Notably, when v_k follows a Beta distribution with parameters 1 and β_0, the joint distribution of $\boldsymbol{\pi}$ corresponds to the Griffiths-Engen-McCloskey (GEM) distribution [66] with a concentration parameter β_0. Furthermore, this stick-breaking process serves as a construction method for the Dirichlet process, a widely utilized nonparametric prior in topic modeling [75].

In our approach, we consider the component weights $\boldsymbol{\pi}$ as the distribution of traversal paths within a document. We postulate that the words within a document originate from a multitude of topic paths. Given the sequential nature of the stick-breaking process, paths with lower serial indices are more inclined to be engaged in representing the document, whereas those with higher serial indices are more likely to remain unengaged. The quantity of engaged paths can be dynamically adjusted by the Stick-Breaking Process (SBP).

Tree-Structured Topic Hierarchy To provide a clearer explanation of our approach, we illustrate by contrasting the sampling procedures across various tree-structured topic models using an example document. As depicted in Fig. 2.12, where node represents a topic z with its distributions over words $\boldsymbol{w}$ and active topics and path are highlighted by boldface, hLDA [4] posits that a document arises from topics located along a single path, which contradicts the assumption underlying topic models that documents can encompass multiple themes. Addressing this limitation, both rCRP [38] and TSNTM [35] propose that any theme within the tree

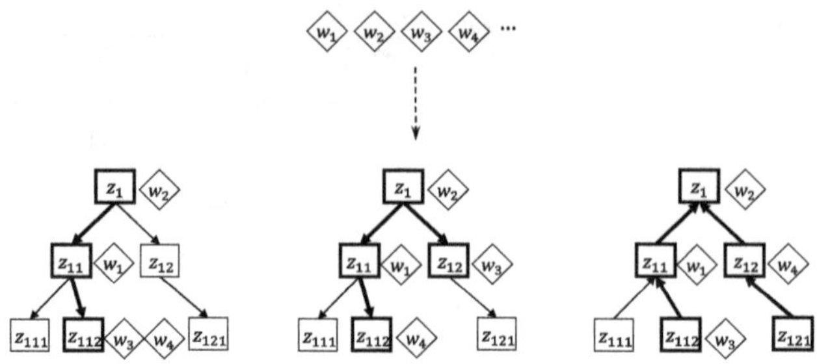

Fig. 2.12 Sampling process of an example document for hLDA [4], rCRP [38], TSNTM [35], and our nTSNTM

Fig. 2.13 Graphical representation of nTSNTM. Solid and dashed arrows denote generation and inference

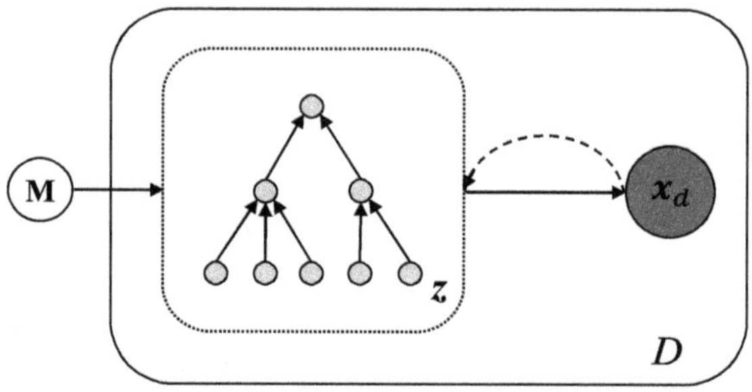

structure can contribute to the generation of a document. Our methodology aligns with the aforementioned perspective adopted by rCRP and TSNTM for modeling hierarchical structures of topics; however, a distinctive aspect is our bottom-up rather than top-down sampling strategy.

Particularly, rCRP employs recursive Chinese Restaurant Process (CRP) starting from the root node to sample topics. On the other hand, TSNTM utilizes a Deep Recurrent Neural Network (DRNN) [1] for path selection from the root, necessitating frequent updates to the tree architecture through heuristic rules. Conversely, our proposed model samples directly from the leaf-level topics, with paths leading back to the root being determined inherently. Specifically, we adopt a conventional stick-breaking construction for inferring the distribution over leaf-level topics, effectively mapping out the distribution of paths. Additionally, dependency matrices are employed to track relationships among different topics, enabling the tree's structural adjustments via backpropagation mechanisms.

As shown in Fig. 2.13, the graphical depiction of the nTSNTM framework is provided. Concerning our methodology, the total number of leaf topics is determined by the Stick-Breaking Process (SBP), whereas the counts of non-leaf topics are regulated through dependency matrices $\mathbf{M}$ spanning different network layers. Specifically, for the lth layer, $\mathbf{M}_l$, an element within $\mathbf{M}$, exists in the space $[0, 1]^{K_l \times K_{l+1}}$, acting as the dependency matrix connecting layers l and $l + 1$. Here, K_l and K_{l+1} denote the maximum possible number of topics at levels l and $l + 1$, respectively. Particularly, $M_{l,k,j}$ represents the likelihood that topic j at level l is the parent of topic k at level $l + 1$, adhering to the condition $\sum_{j'} M_{l,k,j'} = 1$. As discussed in the earlier literature [4], a clearly defined hierarchical structure implies each subtopic is associated with no more than one super-topic. Hence, a softmax function with a low temperature setting [28] is utilized to ensure $\mathbf{M}_{l,k}$ approximates a discrete one-hot vector. This mechanism facilitates the assembly of the topic tree from bottom-up using the introduced $\mathbf{M}$. Additionally, the hierarchy of topics can be dynamically updated in response to modifications in $\mathbf{M}$.

After determining the topic hierarchy by $\mathbf{M}$, the generative process of each word in nTSNTM can be described as follows:

1. For each document $\boldsymbol{x}_d \in \{\boldsymbol{x}_1, \ldots, \boldsymbol{x}_D\}$:

$$\text{Draw SBP weights: } \boldsymbol{\pi}_d \sim \text{GEM}(\boldsymbol{\beta}_0); \tag{2.17}$$

$$\text{Draw Gaussian samples: } \boldsymbol{g}_d \sim \mathcal{N}(\mathbf{0}, \boldsymbol{I}^2); \tag{2.18}$$

$$\text{Draw level distributions: } \boldsymbol{\eta}_d = f_\eta(\boldsymbol{g}_d). \tag{2.19}$$

2. For each word $w_{d,n} \in \{w_{d,1}, \ldots, w_{d,N_d}\}$ in $\boldsymbol{x}_d$:

$$\text{Draw a path: } c_{d,n} \sim \text{Multi}(\boldsymbol{\pi}_d); \tag{2.20}$$

$$\text{Draw a level: } r_{d,n} \sim \text{Multi}(\boldsymbol{\eta}_d); \tag{2.21}$$

$$\text{Draw a word: } w_{d,n} \sim \text{Multi}(\boldsymbol{\phi}_{c_{d,n}[r_{d,n}]}). \tag{2.22}$$

In the above, D is the number of documents, and N_d is the number of words in $\boldsymbol{x}_d$. $\boldsymbol{\phi}_{c_{d,n}[r_{d,n}]} \in \Delta^{V-1}$ is the word distribution of the topic at level $r_{d,n}$ of path $c_{d,n}$, and V is the vocabulary size. $f_\eta(\cdot)$ is a neural perceptron with softmax activation to transform a Gaussian sample to a level distribution.

Parameter Inference Since the Beta distribution does not possess a differentiable non-centered parametrization required by NVI [41], we select the Kumaraswamy distribution [42] to approximate $\text{GEM}(\boldsymbol{\beta}_0)$, i.e., the combination of $\text{Beta}(1, \boldsymbol{\beta}_0)$ with a stick-breaking operation [60]. For the Kumaraswamy distribution, its probability density function on the unit interval is defined as $\text{Kumaraswamy}(x; a, b) = abx^{a-1}(1-x^a)^{b-1}$ for $x \in (0, 1)$ and $a, b > 0$. Sampling from this distribution can be achieved through the inverse transform sampling: $x \sim (1 - u^{\frac{1}{b}})^{\frac{1}{a}}$, where $u \sim \text{Uniform}(0, 1)$. Thus, the KL divergence between the Kumaraswamy distribution and the Beta distribution can be effectively approximated in closed form. Below, we outline the parameter inference process for our nTSNTM.

Firstly, we estimate the component weights of document $\boldsymbol{x}_d$, i.e., $\hat{\boldsymbol{\pi}}_d$, by the following stick-breaking operation with fractions $\boldsymbol{v}_d$:

$$\boldsymbol{\alpha}_d = f_\alpha(\boldsymbol{x}_d), \quad \boldsymbol{\beta}_d = f_\beta(\boldsymbol{x}_d), \tag{2.23}$$

$$\boldsymbol{v}_d \sim \text{Kumaraswamy}(\boldsymbol{\alpha}_d, \boldsymbol{\beta}_d), \tag{2.24}$$

$$\hat{\boldsymbol{\pi}}_d = (v_{d,1}, \ldots, \prod_{j=1}^{T-1} (1 - v_{d,j})), \tag{2.25}$$

where the bag-of-words representation is used for $\boldsymbol{x}_d$. To ensure positive outputs, $f_\alpha(\cdot)$ and $f_\beta(\cdot)$ are neural perceptrons with softplus activation.

Secondly, we infer the level distributions $\hat{\boldsymbol{\eta}}_d$ by

$$\boldsymbol{\mu}_d = f_\mu(\boldsymbol{x}_d), \qquad \boldsymbol{\sigma}_d = f_\sigma(\boldsymbol{x}_d), \tag{2.26}$$

$$\hat{\boldsymbol{g}}_d \sim \mathcal{N}(\boldsymbol{\mu}_d, \boldsymbol{\sigma}_d^2), \quad \hat{\boldsymbol{\eta}}_d = f_\eta(\hat{\boldsymbol{g}}_d), \tag{2.27}$$

where $f_\mu(\cdot)$ and $f_\sigma(\cdot)$ are linear transformations. In practice, we reparameterize $\hat{\boldsymbol{g}}_d = \boldsymbol{\mu}_d + \hat{\epsilon} * \boldsymbol{\sigma}_d$ with the sample $\hat{\boldsymbol{\epsilon}} \sim \mathcal{N}(\boldsymbol{0}, \boldsymbol{I}^2)$ [68].

Thirdly, we obtain the topic distributions of $\boldsymbol{x}_d$, i.e., $\hat{\boldsymbol{\theta}}_d = \{\hat{\boldsymbol{\theta}}_{d,1}, \ldots, \hat{\boldsymbol{\theta}}_{d,L}\}$ by

$$\hat{\boldsymbol{\theta}}_{d,l} = \begin{cases} \hat{\eta}_{d,L}\hat{\boldsymbol{\pi}}_d & \text{if} \quad l = L, \\ \hat{\eta}_{d,l}\hat{\boldsymbol{\pi}}_d \prod_{l' \geq l} \mathbf{M}_{l'} & \text{for} \quad l < L, \end{cases} \tag{2.28}$$

where L denotes the depth of the topic tree, and $\sum_l \sum_k \hat{\theta}_{d,l,k} = 1$.

Subsequently, adhering to the methodology proposed in a previous work [54], we explicitly model the topic-word distributions as $\boldsymbol{\phi} = \text{softmax}(\boldsymbol{u} \cdot \boldsymbol{t}^T)$, where $\boldsymbol{u} \in \mathbb{R}^{V \times H}$ and $\boldsymbol{t} \in \mathbb{R}^{\sum_l K_l \times H}$ denote word vectors and topic vectors, respectively, and H represents the dimension of these vectors.

Given the topic-word distributions $\boldsymbol{\phi}$ and the document-specific topic distributions $\hat{\boldsymbol{\theta}}_d$ obtained from Eq. (2.28), our model reconstructs each document $\boldsymbol{x}_d$ using

$$p(\boldsymbol{w}_{d,n}|\boldsymbol{\phi}, \hat{\boldsymbol{\theta}}_d) = \sum_{z_{d,n}} p(\boldsymbol{w}_{d,n}|\boldsymbol{\phi}_{z_{d,n}}) p(z_{d,n}|\hat{\boldsymbol{\theta}}_d) = \hat{\boldsymbol{\theta}}_d \cdot \boldsymbol{\phi}, \tag{2.29}$$

where $z_{d,n}$ denotes the topic assignment for the word $\boldsymbol{w}_{d,n}$. This formulation allows for a clear probabilistic association between every word in a document and its related topics, based on the learned distributions $\boldsymbol{\phi}$ and $\hat{\boldsymbol{\theta}}_d$, thereby facilitating an efficient reconstruction of the document.

Finally, the variational lower bound of $\boldsymbol{x}_d$ is

$$\begin{aligned} \mathcal{L} =& \mathbb{E}_{q(\boldsymbol{\pi}_d, \boldsymbol{\eta}_d|\boldsymbol{x}_d)}[\sum_n \log(p(w_{d,n}|\boldsymbol{\phi}, \hat{\boldsymbol{\theta}}_d))] \\ &- D_{KL}[q(\boldsymbol{\pi}_d|\boldsymbol{x}_d)||p(\boldsymbol{\pi}_d)] \\ &- D_{KL}[q(\boldsymbol{\eta}_d|\boldsymbol{x}_d)||p(\boldsymbol{\eta}_d)], \end{aligned} \tag{2.30}$$

where $q(\pi_d|\boldsymbol{x}_d)$ and $q(\boldsymbol{\eta}_d|\boldsymbol{x}_d)$ are posteriors modeled by the inference network. $p(\pi_d)$ is the prior for π_d, i.e., GEM($\boldsymbol{\beta}_0$), and $p(\boldsymbol{\eta})$ is the prior for $\boldsymbol{\eta}$, i.e., the standard Gaussian transformed by $f_\eta(\cdot)$.

The parameter inference method for nTSNTM is presented in Algorithm 3. We utilize the variational lower bound to compute gradients and apply the Adam optimizer [39] to update parameters.

Algorithm 3: Parameter inference algorithm

Input: GEM priors $\boldsymbol{\beta}_0$ and documents $\{\boldsymbol{x}_1, \ldots, \boldsymbol{x}_D\}$;
Output: Document-topic distribution $\boldsymbol{\theta}$, topic-word distribution $\boldsymbol{\phi}$, and topic tree $\boldsymbol{Tr}$.
Randomly initialize dependency matrices $\mathbf{M}$ and topic-word distribution $\boldsymbol{\phi}$;
repeat
  for *document* $\boldsymbol{x}_d \in \{\boldsymbol{x}_1, \ldots, \boldsymbol{x}_D\}$ **do**
    Estimate $\hat{\pi}_d$ and $\hat{\eta}_d$ by Eqs. (2.23–2.27);
    Compute $\hat{\boldsymbol{\theta}}_d$ by Eq. (2.28);
    for $w_{d,n} \in \boldsymbol{x}_d$ **do**
      $p(w_{d,n}|\hat{\boldsymbol{\theta}}_d, \boldsymbol{\phi}) = \hat{\boldsymbol{\theta}}_d * \boldsymbol{\phi}$;
    Compute $\mathcal{L}$ by Eq. (2.30);
    Update $f_\alpha(\cdot)$, $f_\beta(\cdot)$, $f_\mu(\cdot)$, $f_\sigma(\cdot)$, $f_\eta(\cdot)$, $\boldsymbol{\phi}$, and $\mathbf{M}$;
until *convergence*;
Build $\boldsymbol{Tr}$ according to $\mathbf{M}$ and $\boldsymbol{\phi}$.

2.2.1.3 Experiments

Datasets We conducted experiments using four widely recognized benchmark datasets: 20NEWS [54], Reuters [85], Wikitext-103 [61], and Rcv1-v2 [54]. The 20NEWS and Reuters datasets consist of news articles, while Wikitext-103 is derived from Wikipedia and used for language modeling tasks. Rcv1-v2 represents an enhanced version of the Reuters dataset. Table 2.7 summarizes the key statistics of these datasets, with the vocabulary compiled using preprocessing steps consistent with the original research. For each dataset, we randomly designated 5% of the training data as the validation set.

Table 2.7 The statistics of datasets

Dataset	#Docs (Train)	#Docs (Test)	Vocabulary size
20NEWS	11,314	7531	1995
Reuters	7769	3019	2000
Wikitext-103	28,472	60	20,000
Rcv1-v2	794,414	10,000	10,000

Experimental Setup For tree-structured topic models, we employ hLDA [4] (also referred to as nCRP [6] in [35]), rCRP [38], and TSNTM [35] as our benchmarks. Following [35], we set the maximum depth of the topic tree to 3 for all these models.

For nonparametric or flat topic models, we utilize HDP [75], GSM and GSB [54], NB-NTM and GNB-NTM [85], and iTM-VAE and HiTM-VAE [62] as our baselines. HDP is a renowned nonparametric topic model capable of accommodating an infinite number of topics. GSM and GSB are two NVI-based models that utilize Gaussian priors, with GSB specifically employing Gaussian distributions to generate stick-breaking fractions. NB-NTM and GNB-NTM are flat neural topic models grounded in Negative Binomial and Gamma Negative Binomial processes, respectively. iTM-VAE and HiTM-VAE extend the work in [60] by incorporating nonparametric processes into the NVI framework, thereby facilitating the extraction of potentially infinite topics.

We leverage publicly available codes for hLDA, rCRP, TSNTM, HDP, NB-NTM, GNB-NTM, iTM-VAE, and HiTM-VAE. We implement GSM and GSB based on the original paper specifications. For parametric models, the number of topics is set to 50 and 200, consistent with [54]. For nonparametric models utilizing the Stick-Breaking Process (SBP), the truncation level is set to 200, and the concentration parameter $\boldsymbol{\beta}_0$ for the GEM distribution is selected from the range [5, 10, 15, 20, 25, 30] using the respective validation set. Specifically, we sequentially select topics whose cumulative probabilities across the entire corpus exceed 95% as active topics. For neural baseline models and our proposed model, we configure the hidden layer size to 256 and use one sample for NVI, following a previous work [54].

All experiments are conducted on a workstation with a Python/Java environment and 40 GB of memory. Notably, we omit the results of hLDA and rCRP on the Rcv1-v2 dataset due to their failure to converge within 48 hours.

Topic Hierarchy Analysis As outlined in [79], a logical topic hierarchy entails that root-level topics are broader, transitioning to more specific ones closer to the leaves. To assess this hierarchy, we utilize topic specialization [38] as a benchmark. Topic specialization is calculated by measuring the cosine distance between a topic's word distribution and the corpus's overall term frequency vector. A higher score signifies a more specialized topic. Figure 2.14 depicts the average specialization scores for each level across various tree-structured models. The findings reveal that nTSNTM and rCRP exhibit a sensible progression of topic specialization across levels, with scores increasing as the hierarchy deepens. Conversely, the TSNTM baseline produces topics at the second level that are more specialized than those at the third, hinting at an illogical hierarchy. In the case of the hLDA baseline, there is a notable jump in specialization from level-2 to level-3, particularly evident in the 20NEWS dataset. This could be attributed to hLDA's mechanism where each document is generated by topics following a single path, leading to a heightened specialization of level-3 topics, as they are confined to a single topic from level-2.

A well-structured topic hierarchy also suggests that child topics maintain coherence with their respective parent topics, as discussed in [79]. To quantify the relationship between two interconnected topics, we have devised a novel metric

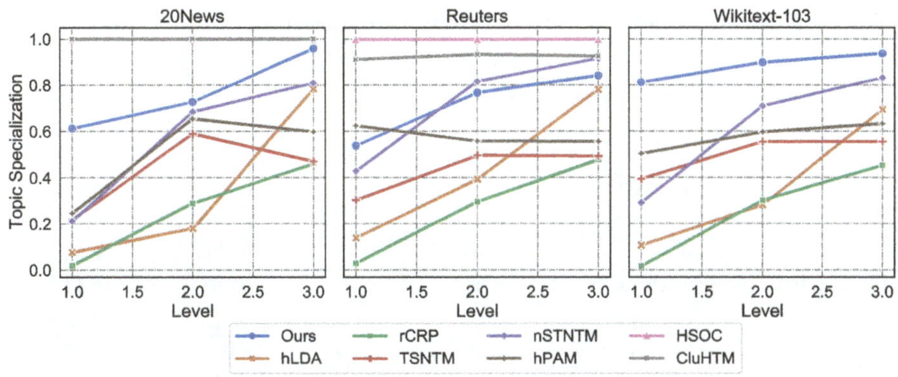

Fig. 2.14 Topic specialization of different tree-structured topic models at each level

known as Cross-Level Normalized Pointwise Mutual Information (CLNPMI). This metric measures the relationship by calculating the average NPMI value between every pair of distinct words, one from the parent topic and one from its child topic. NPMI, introduced by [43], evaluates the relationship between two words w_i and w_j as follows:

$$\text{NPMI}(w_i, w_j) = \frac{\log \frac{P(w_i, w_j)}{P(w_i)P(w_j)}}{-\log(P(w_i, w_j))}. \tag{2.31}$$

Based on NPMI, we define CLNPMI as

$$\text{CLNPMI}(W_p, W_c) = \frac{1}{|W'_p||W'_c|} \sum_{w_i \in W'_p} \sum_{w_j \in W'_c} \text{NPMI}(w_i, w_j), \tag{2.32}$$

where $W'_p = W_p - W_c$ and $W'_c = W_c - W_p$, in which W_p and W_c denote the top-N words of a parent topic and one of its children. To avoid degenerating into NPMI when the parent and the child topics are highly similar, CLNPMI is estimated by the distinct words between every two topics.

To evaluate the topic redundancy for a tree, we introduce a new measurement named the averaged overlap rate (OR) and adopt the widely used topic uniqueness (TU) [61]. OR measures the averaged repetition ratio of top-N words between parent topics and their children, which is defined as $\text{OR}(W_p, W_c) = \frac{|W_p \cap W_c|}{N}$. TU calculates the uniqueness of all topics by $\text{TU} = \frac{1}{K}\sum_{k=1}^{K} \text{TU}(k)$, where K is the number of topics and $\text{TU}(k)$ is defined as

$$\text{TU}(k) = \frac{1}{N}\sum_{n=1}^{N} \frac{1}{cnt(n, k)}. \tag{2.33}$$

Table 2.8 CLNPMI, TU, and OR scores of tree-structured topic models

Model		hLDA	rCRP	TSNTM	nTSNTM
20NEWS	CLNPMI (↑)	0.065	0.098	0.086	**0.122**
	TU (↑)	0.051	0.285	0.430	**0.760**
	OR (↓)	0.056	0.404	0.083	**0.053**
Reuters	CLNPMI (↑)	0.050	0.072	0.027	**0.102**
	TU (↑)	0.447	0.227	0.370	**0.708**
	OR (↓)	0.105	0.515	0.176	**0.066**
Wikitext-103	CLNPMI (↑)	0.063	0.088	0.065	**0.113**
	TU (↑)	0.597	0.355	0.615	**0.730**
	OR (↓)	0.087	0.447	0.078	**0.069**
Rcv1-v2	CLNPMI (↑)	–	–	0.028	**0.088**
	TU (↑)	–	–	0.544	**0.802**
	OR (↓)	–	–	0.051	**0.042**

In the above, $cnt(n, k)$ is the total number of times the nth top word in topic k appears in the top-N words across all topics.

For each of the aforementioned metrics, we compute the average scores considering the top-5, -10, and -15 words. Table 2.8 presents the performance of various models, where each method is executed five times, and the averaged values are reported. Besides, the best value on each metric is highlighted by boldface. The results reveal that our model generally surpasses the baseline models, with p-values below 0.05 indicating statistical significance. However, for the 20NEWS dataset using hLDA and our nTSNTM, the difference in performance based on the OR metric is not statistically significant, with a p-value of 0.391. This underscores the efficacy of nTSNTM's bottom-up structure, where non-leaf topics are activated only when their descendant topics are selected.

We also assess the hierarchical affinity [38] for each model to determine if a parent topic is more similar to its child topics compared to the descendants of other parent topics. Figure 2.15 displays the average cosine similarities between a parent topic's word distribution and both its children topics and non-children topics. For parent topics, both rCRP and nTSNTM exhibit stronger affinities with their children topics than with non-children topics. However, rCRP is plagued by high redundancy, evident from the high similarities (ranging from 0.73 to 0.82) between parent topics and their subtopics.

To visually demonstrate our model's capability in generating a topic tree, we present several topics extracted from the 20NEWS dataset using our nTSNTM and the existing NVI-based TSNTM in Figs. 2.16 and 2.17, respectively. The results indicate that our model can learn a coherent tree-structured topic hierarchy with minimal redundancy. In contrast, TSNTM shows limited discrimination between topics at the second and third levels. Furthermore, topics within the same group at the third level are highly repetitive, such as "rec.sport.baseball" and "talk.politics.misc."

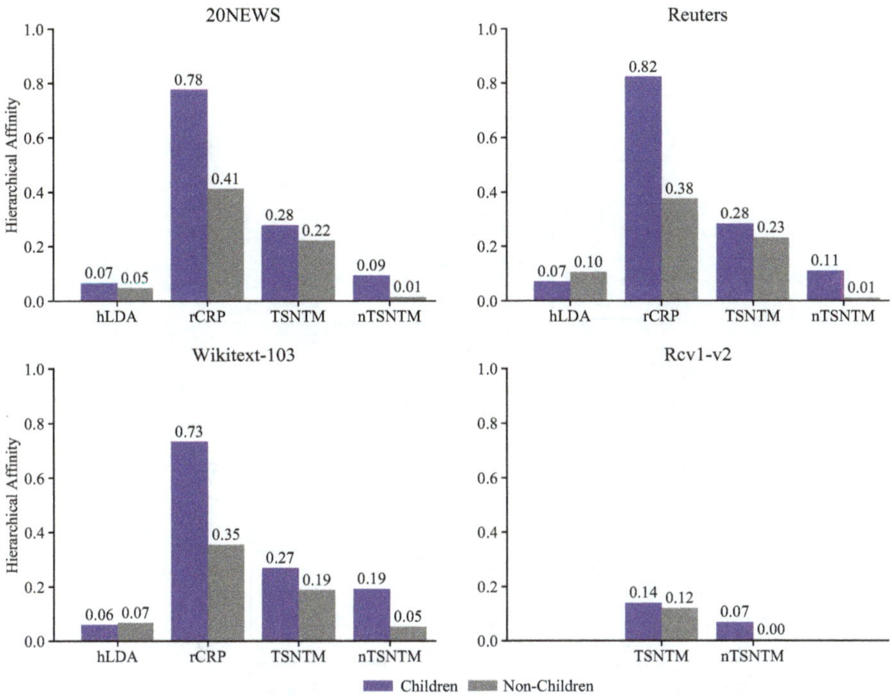

Fig. 2.15 Hierarchical affinity scores

For completeness, we also examine topics extracted from the 20NEWS dataset using hLDA and rCRP. The results reveal that topics at the second level in hLDA are too general to represent distinct topic branches, and the affiliations are unclear. Although rCRP can generate meaningful topics with appropriate affiliations between different levels, it suffers from significant topic redundancy.

Comparison on Topic Interpretability In this section, we utilize the widely adopted NPMI [35, 48, 54, 62, 85] to assess topic interpretability.[3] As noted in [43], NPMI is a metric of topic coherence that aligns closely with the ranking of topic interpretability by human annotators. Table 2.9 displays the NPMI scores for 50 and 200 topics for parametric topic models and topics automatically induced for nonparametric topic models. Each model is run five times, and the averaged results are presented.

Firstly, nTSNTM surpasses all tree-structured baseline models, with the difference being statistically significant at the 0.05 level (except for TSNTM on the

[3] We refrain from estimating perplexity due to two reasons. Firstly, comparing the perplexity of sampling-based and NVI-based models directly is challenging [35]. Secondly, the prior of NVI-based methods significantly influences the perplexity, as the KL divergence may vary greatly across different priors [10].

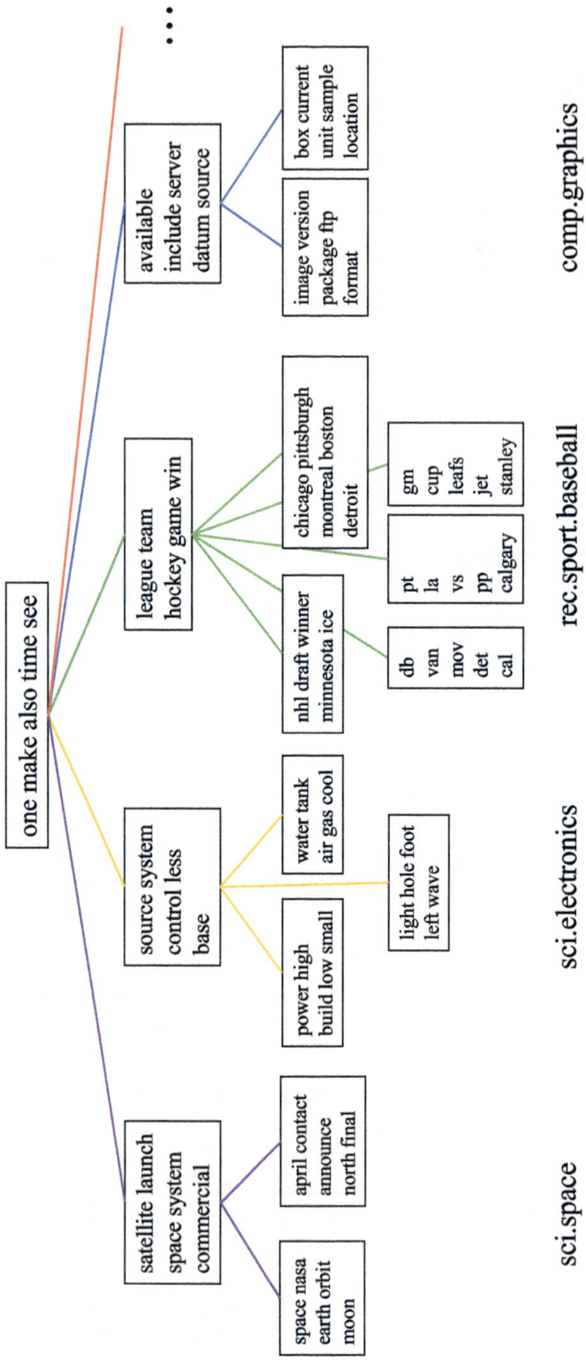

Fig. 2.16 Topic samples extracted from 20NEWS by nTSNTM, where top-5 words are listed for each topic

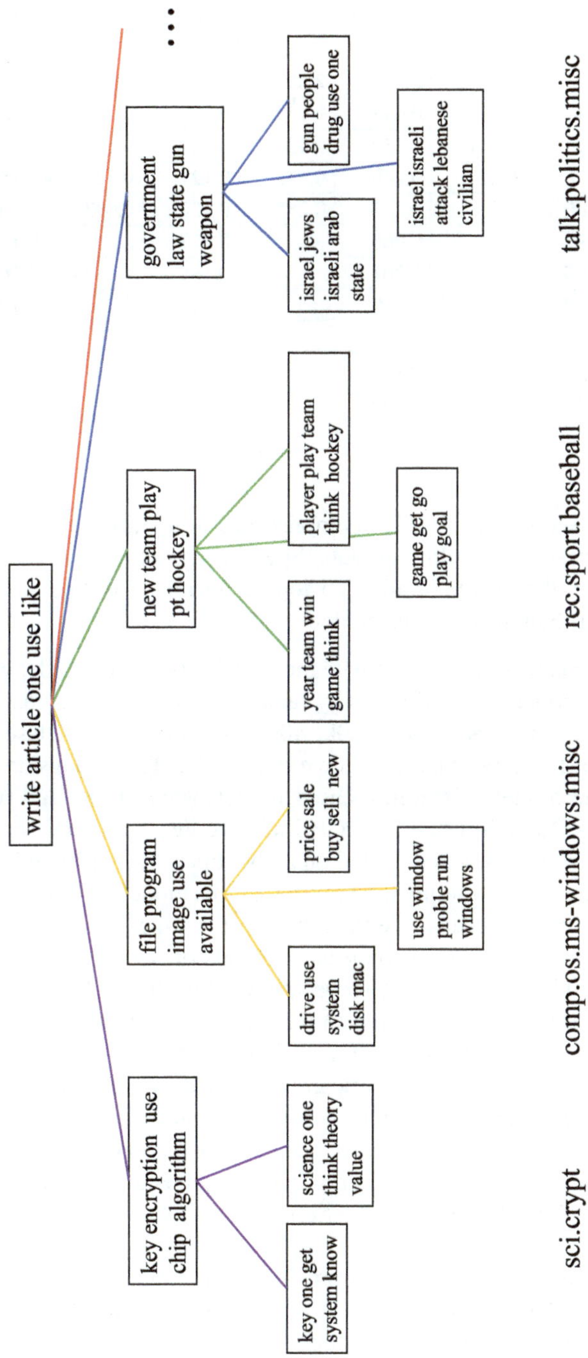

Fig. 2.17 Topic samples extracted from 20NEWS by TSNTM, where top-5 words are listed for each topic

Table 2.9 NPMI of each model, where the best result is marked in bold

	20NEWS NPMI (↑)		Reuters NPMI (↑)		Wikitext-103 NPMI (↑)		Rcv1-v2 NPMI (↑)	
Model	50	200	50	200	50	200	50	200
GSM	0.211	0.165	0.198	0.155	0.214	0.217	**0.231**	0.062
GSB	0.231	0.191	0.152	0.136	0.229	0.131	0.226	0.121
NB-NTM	0.188	0.223	0.248	0.245	0.127	0.125	0.151	0.187
GNB-NTM	**0.240**	0.228	0.237	0.255	0.127	0.093	0.163	0.191
HDP	0.192		0.266		0.157		0.178	
iTM-VAE	0.195		0.201		0.184		0.161	
HiTM-VAE	0.237		**0.269**		0.233		0.179	
rCRP	0.186		0.206		0.201		–	
hLDA	0.221		0.185		0.186		–	
TSNTM	0.212		0.206		0.213		0.225	
nTSNTM	0.237		0.234		**0.237**		0.224	

Rcv1-v2 dataset). Secondly, nTSNTM demonstrates competitive performance when compared to the best flat baseline models. Specifically, except for HiTM-VAE on the Reuters dataset, the results of all other top-performing baseline models are not significantly better than those of our model.

Evaluating Data Scalability To evaluate the scalability of our model with respect to data size, we conducted experiments by randomly sampling various numbers of documents (12.5k, 25k, 50k, 100k, 200k, 400k, and the entire dataset) from the training set of Rcv1-v2. We then ran our model, nTSNTM, alongside other tree-structured baseline models. Sampling-based models (i.e., hLDA and rCRP) were executed on an Intel Xeon Skylake 6133 CPU with 8 cores, while NVI-based models (i.e., TSNTM and nTSNTM) were tested on an Nvidia Tesla V100 GPU.

Figure 2.18 presents the training time for these topic models. The curves of hLDA and rCRP are incomplete because the time they cost is not comparable. Particularly, it costs over 48 hours for training when the numbers of documents are larger than 50k and 100k for hLDA and rCRP, respectively. Our nTSNTM demonstrates an advantage in data scalability compared to the baseline models. Although TSNTM also benefits from GPU acceleration for scalability to large corpora, it employs a doubly recurrent network that significantly slows down its training speed. On the other hand, hLDA and rCRP incur substantial computation time due to path sampling, an issue that becomes more pronounced with larger datasets. Additionally, these two sampling-based models are serial in nature, meaning they can only utilize one core of the CPU, further limiting their efficiency.

Impact of the Concentration Parameter To further validate the nonparametric property of our model, we conducted an analysis to examine the impact of the parameter $\boldsymbol{\beta}_0$ on the number of active topics. Figure 2.19 presents the results of this analysis.

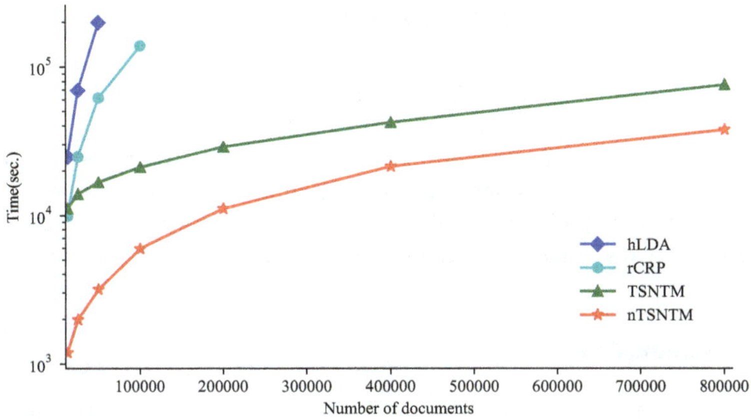

Fig. 2.18 Training time comparison across models along various numbers of documents

Fig. 2.19 Active topic numbers of different models with various values of $\boldsymbol{\beta}_0$

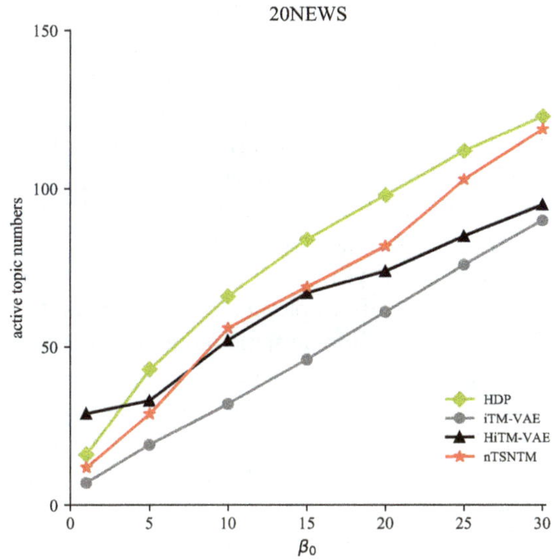

Firstly, we observe that the number of topics increases for all models as we increase the value of $\boldsymbol{\beta}_0$. This is because $\boldsymbol{\beta}_0$ controls the smoothness of the Stick-Breaking Process (SBP). A larger value of $\boldsymbol{\beta}_0$ leads to a smoother distribution, which in turn results in the model identifying more topics.

Secondly, when comparing nTSNTM with iTM-VAE and HiTM-VAE, we find that the number of topics identified by nTSNTM is closer to the number extracted by Hierarchical Dirichlet Process (HDP), a well-known nonparametric topic model. This similarity demonstrates that our model, nTSNTM, is able to approximate the nonparametric property of HDP more effectively than the other compared models.

2.2.2 *Forest-Structured Neural Topic Model*

Building upon the work on nTSNTM, this research reveals that the actual structure of topics in real-world data tends to form a forest rather than a single tree. It introduces the nFNTM, a novel nonparametric **forest-structured** neural topic model. This model employs **self-attention mechanisms** to identify relationships between parent and child topics and constructs a sparse **Directed Acyclic Graph (DAG)** to effectively organize topics into a comprehensive forest structure.

2.2.2.1 Introduction

Topic modeling is a widely used technique for representing a collection of documents and encoding their text content into a lower dimensional feature space. Traditional topic models can broadly be categorized into probabilistic graphical models and matrix factorization-based methods. Probabilistic graphical models, such as Latent Dirichlet Allocation (LDA) [5], are reliant on approximate methods like variational inference and Gibbs sampling for parameter estimation. These methods often involve complex derivations and can be computationally expensive [73]. On the other hand, matrix factorization-based methods, such as those proposed by Lee and Seung [44], effectively decompose the document-word matrix into two smaller matrices. However, these methods can suffer from low stability in their results [15]. Recently, neural topic models, which leverage the Neural Variational Inference (NVI) [15, 54, 73], have gained significant attention. These models offer the advantages of fast parameter inference and flexibility, making them a promising alternative to traditional topic modeling approaches.

Despite the notable successes of neural topic models, a significant limitation of many existing methods is their inability to capture hierarchical structures beyond a single-root tree. This can lead to difficulties in identifying the relationships among topics, which is crucial for data analysis and exploration across various domains [64].

To address this issue, several neural topic models have been developed that can model hierarchical topic structures [14, 15, 35]. However, these methods all assume a hierarchical structure in the form of a tree with a single-root node. This assumption is a significant limitation, as real-world corpora often exhibit more complex hierarchical structures.

For example, in a corpus of emails and academic studies, topics may be organized into several independent trees. As illustrated in Fig. 2.20, there could be two root topics: one related to emails and another related to study. Under these root topics, there are various second-level topics, such as the contents of emails (politics and religion) and modes of study (patients and rats). Some of these second-level topics have multiple child topics that delve deeper into specific areas (e.g., country and community within politics, or cell, protein, and genes within biology), while others may have only one child topic or none. This example demonstrates that the real-

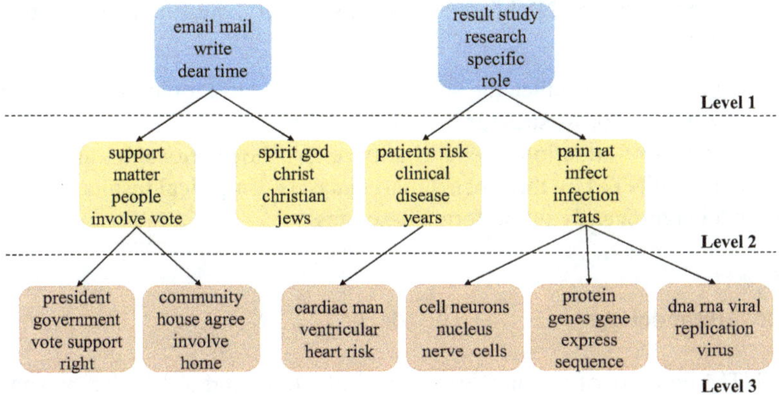

Fig. 2.20 An example of forest-structured topics. Each topic is represented by top-5 words

world hierarchical topic structure is more accurately represented as a forest of trees rather than a single tree. Therefore, there is a need for neural topic models that can capture this more complex hierarchical structure.

The implementation of forest-structured topic models continues to encounter obstacles, particularly due to the necessity for their hierarchical topic arrangement to adhere to several principles. Firstly, the structure must exhibit coherence, where root topics encapsulate general ideas, while child topics specialize in the contexts of their respective parent topics, as emphasized in [79]. Secondly, topics should demonstrate affinity, meaning each topic exhibits greater similarity to its own offspring compared to topics descended from different parents, as highlighted in [38]. Thirdly, diversity is crucial; the distributions of words associated with parent and child topics should be discernibly distinct, as noted in [4]. Lastly, flexibility is essential; children of each topic should be dynamically assigned [38], and the number of topics at each hierarchical level should be unrestricted (i.e., nonparametric), as suggested in [15]. Furthermore, predefining the number of root topics, as well as the depth and breadth of the forest structure, poses additional difficulties.

In this work, we introduce the nonparametric Forest-structured Neural Topic Model (nFNTM) as a solution to these challenges. This model initially captures parent-child topic relationships through the utilization of the self-attention mechanism, as cited in [78].[4] Subsequently, the model learns a sparse Directed Acyclic Graph (DAG) to construct a topic forest by integrating document-topic distributions and parent-child relationships using Noise-Contrastive Estimation (NVI). To our knowledge, existing DAG-structured topic models primarily rely on Bayesian learning [46, 55] or Nonnegative Matrix Factorization (NMF) [49, 79], with no

[4] We opt for the self-attention mechanism because it efficiently captures global dependencies among input tokens, minimizing dependence on external information, as highlighted in [91]. Additionally, it facilitates parallel processing for enhanced computational efficiency.

prior work employing the NVI framework. In summary, our main contributions are outlined as follows:

- We are the first to introduce the sparse DAG into neural topic modeling with the aim of learning a forest-structured topic hierarchy.
- We develop a self-attention mechanism to capture the relationships among topics.
- We evaluate nFNTM on three benchmark datasets. Empirical results indicate that our model significantly outperforms baselines.

2.2.2.2 Methodology

Our nFNTM consists of an encoder, a topic attention, and a decoder, as shown in Fig. 2.21.

Encoder Given a collection of documents, each document $\boldsymbol{d} \in \mathbb{R}^V$ is represented by bag-of-words (BoW), where V denotes the vocabulary size. For the encoder, we transform $\boldsymbol{d}$ into a document-topic distribution $\boldsymbol{\pi}$ using the Stick-Breaking Process (SBP) [34]. SBP provides a mechanism for defining atomic measures in Bayesian nonparametric methods. An almost sure (a.s.) discrete probability measure $\mathcal{P}$ can be considered an SBP if it can be expressed as

$$\mathcal{P} = \sum_{i=1}^{\infty} \pi_i \delta_{x_i}, \quad \pi_i = \begin{cases} v_1 & \text{if } i = 1, \\ v_i \prod_{t<i} (1 - v_t) & \text{if } i > 1, \end{cases} \tag{2.34}$$

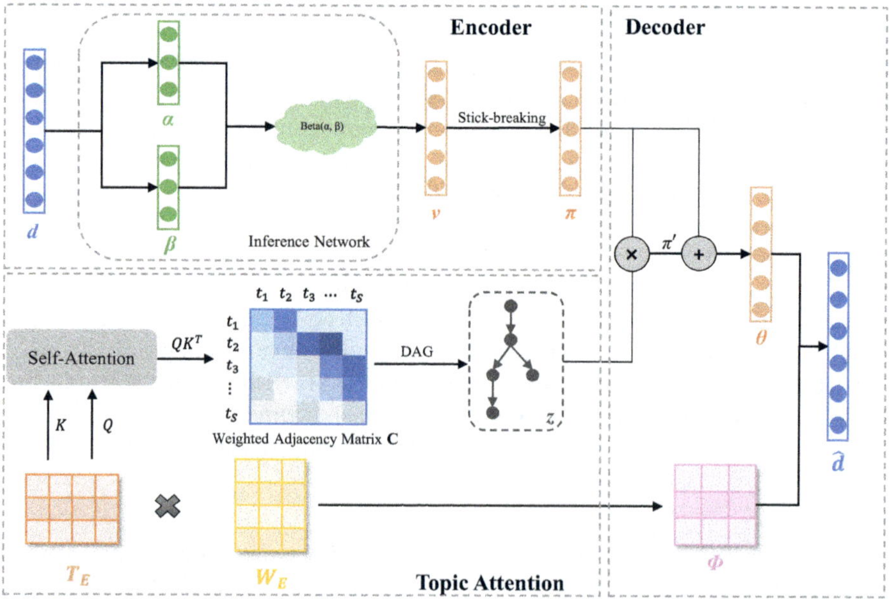

Fig. 2.21 Structure of nFNTM

where $x_i \sim \mathcal{H}$, $\mathcal{H}$ is the base probability measure, δ_{x_i} is a discrete measure concentrated at x_i, and $\boldsymbol{v} \sim \text{Beta}(\boldsymbol{\alpha}, \boldsymbol{\beta})$, with $\boldsymbol{\alpha}$ and $\boldsymbol{\beta}$ being prior parameters. The random weights $\{\pi_i\}$ are independent of $\mathcal{H}$ and satisfy $0 \leq \pi_i \leq 1$ and $\sum_{i=1}^{\infty} \pi_i = 1$. When $\boldsymbol{v} \sim \text{Beta}(1, \hat{\boldsymbol{\beta}})$, the SBP corresponds to the Dirichlet process, and the joint distribution over the infinite sequence of stick-breaking weights with concentration parameter $\hat{\boldsymbol{\beta}}$ is $\{\pi_i\} \sim \text{GEM}(\hat{\boldsymbol{\beta}})$ [75].

Notably, nTSNTM [15] uses the Kumaraswamy distribution [42] to approximate the Beta distribution due to its differentiability properties. However, the Beta distribution is part of a one-parameter family of symmetric distributions that offers various ways to generate distributions via physical processes [37]. To estimate the Beta distribution without bias, we employ implicit reparameterization gradients, which will be detailed in Sect. 2.2.2.2. As shown in Fig. 2.21, we introduce an inference network to construct the Beta distribution. We derive $\boldsymbol{\alpha}$ and $\boldsymbol{\beta}$ through $\boldsymbol{\alpha} = l_\alpha(\boldsymbol{\eta})$ and $\boldsymbol{\beta} = l_\beta(\boldsymbol{\eta})$, where $\boldsymbol{\eta} = \text{MLP}(\boldsymbol{d})$ represents the low-dimensional representation of $\boldsymbol{d}$, l_α and l_β denote single linear layers, and MLP indicates a multilayer perceptron.

Topic Attention To enable the model to focus on relationships among topics in a way that distributes attention according to the corpus-generated topics, we utilize topic embeddings $\mathbf{T_E} \in \mathbb{R}^{S \times d_t}$ for performing attention operations on topics. Here, $S \to \infty$ represents the breaking number in the Stick-Breaking Process (SBP), which corresponds to the number of topics learned by SBP, and d_t denotes the dimension of topic embeddings. Considering that each topic's relevance should be computed and learned independently, we treat each topic embedding as a separate subspace and project input hidden representations into these different topic embedding subspaces as follows:

$$\mathbf{Q} = \mathbf{T_E}\mathbf{W_Q}, \quad \mathbf{K} = \mathbf{T_E}\mathbf{W_K}, \tag{2.35}$$

where $\mathbf{W_Q} \in \mathbb{R}^{d_t \times d_r}$ and $\mathbf{W_K} \in \mathbb{R}^{d_t \times d_r}$ are trainable parameters, d_r is the dimension of each topic embedding subspace, and $\mathbf{Q} \in \mathbb{R}^{S \times d_r}$ and $\mathbf{K} \in \mathbb{R}^{S \times d_r}$ denote the matrices of queries and keys. Based on these queries and keys, we compute a weighted adjacency matrix $\mathbf{C} \in \mathbb{R}^{S \times S}$, where each element $C_{i,j}$ reflects the parent-child correlation degree between topics t_i and t_j.[5] To ensure that each element $C_{i,j}$ is discrete, a softmax function with temperature τ is applied [28]:

$$C_{i,j} = \frac{\exp(\frac{\hat{C}_{i,j}}{\tau})}{\sum_{k=1}^{S} \exp(\frac{\hat{C}_{i,k}}{\tau})}, \tag{2.36}$$

where $\hat{\mathbf{C}} = \mathbf{Q}\mathbf{K}^T$.

[5] Different from [78], we capture the relationships among topics without using the value matrix. This approach was chosen because our experiments showed that using a value matrix could lead to all topics converging to a few important ones.

To integrate parent-child correlations into the document-topic distributions while preserving the nonparametric nature of SBP, we initially calculate the parent document-topic distribution by integrating the weighted adjacency matrix, i.e., $\boldsymbol{\pi}' = \boldsymbol{\pi} \times \mathbf{C}$. Initially, since the weighted adjacency matrix does not yet reflect parent-child correlations accurately, the document-topic distribution is represented by $\boldsymbol{\pi}$. As the learning of our self-attention module progresses, the quality of the weighted adjacency matrix $\mathbf{C}$ improves, making $\boldsymbol{\pi}'$ increasingly valuable due to stronger parent-child topic relationships. We then generate the consolidated document-topic distribution by combining both $\boldsymbol{\pi}'$ and $\boldsymbol{\pi}$ as $\boldsymbol{\theta} = (1 - \gamma)\boldsymbol{\pi}' + \gamma\boldsymbol{\pi}$, where γ serves as a decay coefficient.

Decoder Given that bag-of-words (BoW) document representations do not include word-relatedness information, we integrate pretrained word embeddings $\mathbf{W_E}$ [79, 85] into our network. For the decoder part, we initially compute the topic-word distribution $\boldsymbol{\Phi}$ using a softmax function over the product of topic embeddings and word embeddings:

$$\boldsymbol{\Phi} = \mathrm{softmax}(\mathbf{T_E} \times \mathbf{W_E}), \tag{2.37}$$

where $\mathbf{W_E} \in \mathbb{R}^{d_t \times V}$ represents the word embeddings matrix with d_t being the dimension of topic embeddings and V the vocabulary size, while $\boldsymbol{\Phi} \in \mathbb{R}^{S \times V}$ denotes the topic-word distribution matrix, with S indicating the number of topics.

To reconstruct the document $\hat{\boldsymbol{d}}$, we combine the document-topic distribution $\boldsymbol{\theta}$ with the topic-word distribution $\boldsymbol{\Phi}$. Specifically, the reconstruction process involves multiplying these two distributions to generate an estimated document representation $\hat{\boldsymbol{d}}$ that incorporates both topic and word-level information:

$$\hat{\boldsymbol{d}} = \boldsymbol{\theta} \times \boldsymbol{\Phi}. \tag{2.38}$$

This approach ensures that the reconstructed document $\hat{\boldsymbol{d}}$ not only reflects the thematic structure captured by the topic model but also preserves semantic relationships between words as encoded in the pretrained word embeddings.

Topic Forest Hierarchy To construct a topic forest, we develop a hierarchical structure of topics based on the weighted adjacency matrix $\mathbf{C}$, which encodes parent-child relationships among topics. However, without constraints, $\mathbf{C}$ might not form a reasonable hierarchical structure and could contain loops. To address this issue, we propose to structure topics as a sparse Directed Acyclic Graph (DAG). Zheng et al. [95] demonstrated that a weighted adjacency matrix $\mathbf{C}$ represents a DAG if and only if $h(\mathbf{C}) = \mathrm{tr}(e^{\mathbf{C} \circ \mathbf{C}}) - d = 0$, where $\circ$ denotes the Hadamard product and $e^{\mathbf{C}}$ is the matrix exponential of $\mathbf{C}$. We enforce $h(\mathbf{C}) = 0$ using an augmented Lagrangian method to ensure the acyclicity of $\mathbf{C}$.

Given that the topic hierarchy is reasonable and child topics can be represented by their parent topics, we hypothesize that the sum of child topics' document-topic distributions closely resembles their parent topics' distributions. Therefore, we learn the forest-structured topic hierarchy by minimizing the discrepancy between documents reconstructed from $\boldsymbol{\pi} \times \boldsymbol{\Phi}$ and those reconstructed from their parent document-topic distributions ($\boldsymbol{\pi}' \times \boldsymbol{\Phi}$), under the constraint $h(\mathbf{C}) = 0$ via the augmented Lagrangian method:

$$\begin{aligned}\min_{\mathbf{C}\in\mathbb{R}^{S\times S}} \quad \mathcal{L}_C = & \frac{1}{2}\|(\boldsymbol{\pi} - \boldsymbol{\pi}') \times \boldsymbol{\Phi}\|_F^2 \\ & + \frac{\rho}{2}|h(\mathbf{C})|^2 + \epsilon h(\mathbf{C}),\end{aligned} \tag{2.39}$$

where ρ is a penalty parameter, and ϵ is the Lagrange multiplier. Parameters ρ and ϵ are updated following [95] as follows:

$$\begin{cases} \rho_i & = 2\rho_{i-1}, \\ \epsilon_i & = \epsilon_{i-1} + \rho h_{i-1}, \end{cases}$$

with initial values $\rho_0 = 1$ and $\epsilon_0 = 0$, where i indicates the training epoch and h is the constraint value.

The generative process of nFNTM is described as follows:

1. For each document $\boldsymbol{d}$:
   a. Draw a topic proportion $\boldsymbol{\pi} \sim \text{GEM}(\hat{\boldsymbol{\beta}})$.
   b. Get the weighted adjacency matrix $\mathbf{C}$.
   c. Get the correlational topic distribution $\boldsymbol{\theta}$.
2. For each word $w_{d,n} \in \boldsymbol{d}$:
   a. Draw a topic $z_{d,n} \sim \text{Mult}(\boldsymbol{\theta})$.
   b. Obtain the topic-word distribution $\boldsymbol{\Phi}$.
   c. Draw a word $w_{d,n} \sim \text{Mult}(\boldsymbol{\Phi}_{z_{d,n}})$.

Parameter Inference We apply NVI to inference network parameters, which is proven to be efficient and flexible [15, 35, 54, 73]. The likelihood of each reconstructed document $\hat{\boldsymbol{d}}$ is estimated by $p\left(\hat{\boldsymbol{d}} \mid \boldsymbol{\theta}, \boldsymbol{\Phi}\right) = \sum_z p\left(\hat{\boldsymbol{d}} \mid \boldsymbol{\Phi}_z\right) p\left(z \mid \boldsymbol{\theta}\right)$, where z is the topic assigned for each word in $\hat{\boldsymbol{d}}$. To maximize the log-likelihood, we derive the lower bound as follows:

$$\begin{aligned}\mathcal{L}_B = & \mathbb{E}_{q(\boldsymbol{\theta},\boldsymbol{\Phi}|\boldsymbol{d})}\left[\log p\left(\hat{\boldsymbol{d}} \mid \boldsymbol{\theta}, \boldsymbol{\Phi}\right)\right] \\ & - D_{KL}\left[q\left(\boldsymbol{\theta} \mid \boldsymbol{d}\right) \| p\left(\boldsymbol{\theta}\right)\right] - \mathcal{L}_C,\end{aligned} \tag{2.40}$$

where $\mathbb{E}_{q(\boldsymbol{\theta},\boldsymbol{\Phi}|\boldsymbol{d})}\left[\log p\left(\hat{\boldsymbol{d}} \mid \boldsymbol{\theta}, \boldsymbol{\Phi}\right)\right]$ is the reconstruction loss, and $D_{KL}\left[q\left(\boldsymbol{\theta} \mid \boldsymbol{d}\right) \| p\left(\boldsymbol{\theta}\right)\right]$ is the Kullback-Leibler (KL) divergence between the prior Beta distribution $p\left(\boldsymbol{\theta}\right)$ and the posterior Beta distribution $q\left(\boldsymbol{\theta} \mid \boldsymbol{d}\right)$. The

KL divergence of two Beta distributions is given below:

$$
\begin{aligned}
&KL(Beta(\alpha_1, \beta_1)||Beta(\alpha_2, \beta_2)) = \\
&\quad \ln\Gamma(\alpha_2) + \ln\Gamma(\beta_2) + \ln\Gamma(\alpha_1 + \beta_1) \\
&\quad - (\ln\Gamma(\alpha_1) + \ln\Gamma(\beta_1) + \ln\Gamma(\alpha_2 + \beta_2)) \\
&\quad + (\alpha_1 - \alpha_2)\digamma(\alpha_1) \\
&\quad + (\beta_1 - \beta_2)\digamma(\beta_1) \\
&\quad + (\alpha_2 + \beta_2 - \alpha_1 - \beta_1)\digamma(\alpha_1 + \beta_1),
\end{aligned}
\tag{2.41}
$$

where $\digamma(x) = \frac{d}{dx}\ln\digamma(x) = \frac{\digamma'(x)}{\digamma(x)}$.

Note that we transform $\boldsymbol{d}$ to the variational Beta distribution; thus $q(\boldsymbol{\theta} \mid \boldsymbol{d})$ is derived by

$$
\begin{aligned}
q(\boldsymbol{\theta} \mid \boldsymbol{d}) &= \mathrm{Beta}(\boldsymbol{\theta} \mid \boldsymbol{\alpha}, \boldsymbol{\beta})) \\
&= \mathrm{Beta}(\boldsymbol{\theta} \mid l_\alpha(\mathrm{MLP}(\boldsymbol{d})), l_\beta(\mathrm{MLP}(\boldsymbol{d}))).
\end{aligned}
\tag{2.42}
$$

To compute reparameterization gradients, the Beta samples are obtained from Gamma samples since the latter do not require inverting the standardization function [23]: For $z_1 \sim \mathrm{Gamma}(\boldsymbol{\alpha}, 1)$ and $z_2 \sim \mathrm{Gamma}(\boldsymbol{\beta}, 1)$, it has $\frac{z_1}{z_1+z_2} \sim \mathrm{Beta}(\boldsymbol{\alpha}, \boldsymbol{\beta})$. The lower bound $\mathcal{L}_B$ is used to calculate gradients, and parameters are updated by Adam [40].

2.2.2.3 Experimental Results

Experimental Setup To thoroughly assess the efficacy of our proposed nFNTM, we undertake experiments on three datasets known for their hierarchical nature, as noted in [15]: 20News [54], Reuters [85], and Wikitext-103 [61]. As per [49, 79], topics often encompass subtopics, forming a hierarchical structure in web directories of news (such as 20News and Reuters) and encyclopedias (like Wikitext-103, derived from Wikipedia). In these datasets, each document is paired with a hierarchy of manually curated categories. For instance, in 20News, "rec.autos" and "rec.motor.cycles" are subcategories of "rec." Instead of relying on predefined, coarse-grained categories, we construct a hierarchical structure from the corpus at a fine-grained topic level. This is achieved by automatically extracting a set of representative words for each topic within the forest structure, aiding users in understanding their topics of interest.

Prior to experimentation, all datasets underwent preprocessing to remove stop words and low-frequency terms. Table 2.10 outlines the number of training and test documents, along with the vocabulary size for each dataset.

Table 2.10 The statistics of datasets

Dataset	#Docs (Train)	#Docs (Test)	Vocabulary size
20News	11,314	7531	1995
Reuters	7769	3019	2000
Wikitext-103	28,472	120	20,000

To benchmark the performance of our nFNTM, we utilize a diverse range of baseline models, encompassing four tree-structured topic models: hLDA [4], rCRP [38], TSNTM [35], and nTSNTM [15]. Additionally, we include three DAG-structured topic models: hPAM [55], HSOC [49], and CluHTM [79]. For the tree-structured baselines, we adhere to the setting in [35] by fixing the maximum depth of the topic tree at 3. Furthermore, we compare against seven flat topic models, including parametric models such as GSM [54], GSB [54], NB-NTM [85], and GNB-NTM [85], as well as nonparametric models like HDP [75], iTM-VAE [62], and HiTM-VAE [62]. We leverage open-source codes for all baseline models, with the exception of GSB.

To ensure fair comparisons, we follow [15] to set topic numbers to 50 and 200 for all parametric models. For nonparametric models based on SBP (i.e., iTM-VAE and nTSNTM), the maximum number of topics is set to 200, and the concentration parameter $\hat{\boldsymbol{\beta}}$ of the GEM distribution is set to 20. According to [15], we select the topic with a total probability exceeding 95% as an active topic. Besides the penalty parameter ρ is updated by $\rho = 2^x$, where x denotes the number of training epochs. The temperature τ changes exponentially from 5 to 1×10^{-4} to ensure the adjacency matrix to be sparse. The hidden layer size of the encoder is set to 256, which is consistent with other models. In the construction of the weighted adjacency matrix, we use an exponential change strategy to dynamically adjust the decay coefficient. This is because the quality of such a weighted adjacency matrix is gradually enhanced when learning the self-attention module. Finally, the decay coefficient γ changes exponentially from 1 to 0.5. We implement our model by PyTorch and run it on a computer with NVIDIA 1080Ti and 128GB RAM.

To maintain fairness in our comparisons, we adhere to the settings outlined in [15], setting the number of topics to 50 and 200 for all parametric models. For nonparametric models based on the Stick-Breaking Process (SBP), specifically iTM-VAE and nTSNTM, we cap the maximum number of topics at 200 and set the concentration parameter $\hat{\boldsymbol{\beta}}$ of the GEM (Griffiths, Engen, and McCloskey) distribution to 20. Following [15], we consider a topic active if its total probability exceeds 95%. Additionally, we update the penalty parameter ρ using an exponential strategy, where $\rho = 2^x$, with x representing the number of training epochs. The temperature τ is adjusted exponentially, decreasing from 5 to 1×10^{-4} to ensure sparsity in the adjacency matrix.

Consistent with other models, we set the hidden layer size of the encoder to 256. In constructing the weighted adjacency matrix, we employ an exponential change strategy to dynamically adjust the decay coefficient. This approach is adopted because the quality of the weighted adjacency matrix improves gradually as the self-attention module learns. Specifically, the decay coefficient γ changes exponentially from 1 to 0.5. Our model is implemented using PyTorch and executed on a computer equipped with an NVIDIA 1080Ti GPU and 128 GB of RAM.

Topic Hierarchy Analysis To evaluate the rationality, affinity, and diversity of the topic hierarchy generated by different models, we adopt four metrics: topic specialization [38], Cross-Level Normalized Pointwise Mutual Information (CLNPMI) [15], hierarchical affinity [38], and Topic Uniqueness (TU) [61]. Key words of each topic are ranked by the topic-word matrix $\mathbf{\Phi}$ [5].

Topic Hierarchical Rationality Regarding the tree-structured topics, topics situated nearer to the root node exhibit a tendency toward greater generality, whereas those nearer to the leaf node demonstrate increased specificity. The Topic Specialization metric quantifies this attribute by determining the cosine similarity between the word distribution of individual topics and the comprehensive corpus. Given that our forest-structured model produces numerous trees, we opt to utilize all three-tier topic trees to compute an average score for comparative purposes.

The Topic Specialization outcomes are displayed in Fig. 2.22. Upon examining this figure, it becomes apparent that our model surpasses the baseline models, with the exception of HSOC and CluHTM. Notably, despite the Topic Specializations of HSOC and CluHTM at various levels approaching 1, they still fall short in rationality, as the root topic in a tree structure ideally should encompass a broader scope compared to its subtopics. Nevertheless, it is crucial to acknowledge that a high specialization score alone does not serve as a definitive indicator of the overall quality of the topic hierarchy; other metrics, such as affinity and diversity, must also be taken into account.

To assess the relationship between interconnected topics, a metric termed CLNPMI was introduced in [15]. This metric calculates the average Normalized Pointwise Mutual Information (NPMI) score across pairs of parent and child topics. The formula for CLNPMI is expressed as follows:

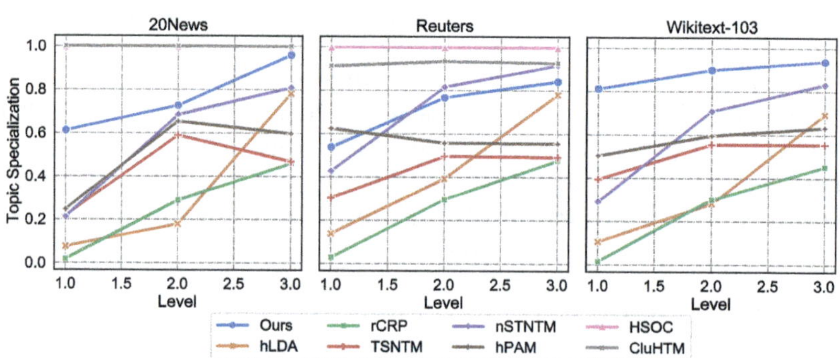

Fig. 2.22 Topic specialization of different topics trees generated from the three datasets

$$\text{CLNPMI}(W_p, W_c) = \frac{1}{|W'_p||W'_c|} \sum_{w_i \in W'_p} \sum_{w_j \in W'_c} \text{NPMI}(w_i, w_j).$$

In the above equation, W'_p and W'_c represent the sets $W_p \setminus W_c$ and $W_c \setminus W_p$, respectively, where W_p and W_c denote the top-N words associated with a parent topic and its corresponding child topic. A higher CLNPMI value indicates a stronger coherence between the child topics and their respective parent topics. In our study, we compared our proposed nFNTM model with various hierarchical topic models, and the CLNPMI results for all models are presented in Table 2.11. It is important to note that two NMF-based models, specifically HSOC and CluHTM, failed to converge within a reasonable time (48 hours) on the Wikitext-103 dataset, and therefore their results were omitted from the table. A higher score means better performance, and the best scores are highlighted by boldface. As shown in Table 2.11, our nFNTM model exhibited the best performance across all datasets.

Topic Uniqueness To quantify the level of diversity within the generated hierarchical topics, we introduce a metric known as Topic Uniqueness (TU). This metric is computed as follows:

$$\text{TU} = \frac{\text{Number of Unique Words in Top } N \text{ Words of All Topics}}{N \times \text{Total Number of Topics}}.$$

A higher TU value signifies that the topics generated are more distinctive and diverse. Table 2.11 presents the TU results for various models, demonstrating that our proposed model yields topics with greater diversity compared to other models. Notably, the baseline models, namely HSOC and CluHTM, perform poorly due to their requirement for a preset number of topics, often as high as 600, to achieve convergence.

Table 2.11 The average CLNPMI and TU scores of hierarchical topic models with top-5, -10, and -15 words for each topic

	20News		Reuters		Wikitext-103	
Models	*CLNPMI*	*TU*	*CLNPMI*	*TU*	*CLNPMI*	*TU*
hLDA	0.065	0.051	0.050	0.447	0.063	0.597
rCRP	0.098	0.285	0.072	0.227	0.088	0.355
TSNTM	0.086	0.430	0.027	0.370	0.065	0.615
nTSNTM	0.109	0.745	0.102	0.708	0.113	0.730
hPAM	0.046	0.606	0.011	0.470	0.047	0.713
HSOC	0.128	0.231	0.047	0.211	–	–
CluHTM	0.123	0.116	0.016	0.117	–	–
Ours	**0.152**	**0.757**	**0.125**	**0.798**	**0.118**	**0.766**

Topic Hierarchical Affinity Assuming that topics with hierarchical parent-child relationships demonstrate a higher degree of similarity in their word distributions compared to those without such relationships, as proposed in [38], we initially evaluate the similarity between parent-child topics and non-parent-child topics by calculating the cosine similarity of their respective word distributions. Following this, we measure the hierarchical affinity of the topics by examining the disparity between these similarities.

Let $\Phi(k)$ represent a topic at level k, $\Lambda(k)$ denote the set of child topics of $\Phi(k)$, and $\overline{\Lambda}(k)$ signify the set of non-child topics of $\Phi(k)$. The hierarchical affinity metric is defined as the comparison between the average cosine similarity $S_{\Lambda}(k)$ between $\Phi(k)$ and all topics in $\Lambda(k)$ and the average cosine similarity $S_{\overline{\Lambda}}(k)$ between $\Phi(k)$ and all topics in $\overline{\Lambda}(k)$. A substantial difference between $S_{\Lambda}(k)$ and $S_{\overline{\Lambda}}(k)$ indicates a strong hierarchical affinity.

Given that all topics between paired levels in the hPAM model are fully connected, it is not feasible to clearly differentiate between topics with parent-child relationships and those without. Therefore, we exclude the hPAM model from this analysis. As illustrated in Fig. 2.23, our proposed model exhibits high similarities between parent and child topics and low similarities between non-parent and child topics, suggesting a robust hierarchical affinity. It is noteworthy that both the HSOC and CluHTM models are based on Nonnegative Matrix Factorization (NMF) and rely on a predefined number of topic trees. To avoid missing potential topic substructures, these models require a significant number of topic trees (e.g., 100) according to their default settings. However, this results in an underestimation of the similarity between non-parent-child topics due to the large number of such topics, leading to a seemingly competitive hierarchical affinity. Nevertheless, as demonstrated in Table 2.11, this approach results in a poor Topic Uniqueness (TU) score for both HSOC and CluHTM.

Topic Interpretability To assess the interpretability of topics, we utilize the Normalized Pointwise Mutual Information (NPMI) score, as it has been demonstrated to align well with human judgments [43]. We extract the top-5, -10, and -15 words for

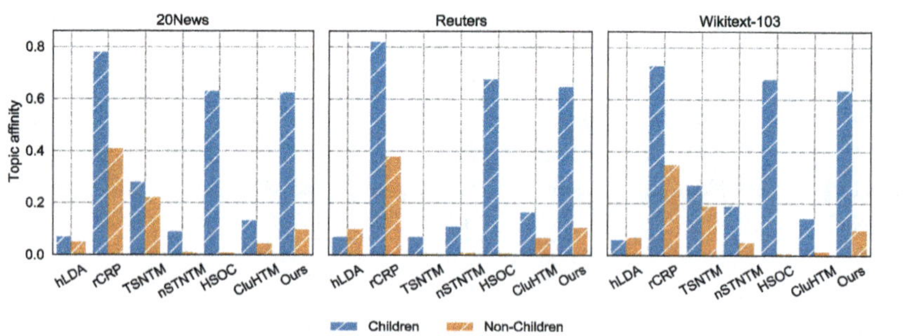

Fig. 2.23 Topic hierarchical affinity results. For each associated topic, a larger difference of word distributions between children and non-children topics is better

Table 2.12 The average NPMI scores of different models using top-5, -10, and -15 words for each topic

Datasets	20News		Reuters		Wikitext-103	
Model	50	200	50	200	50	200
GSM	0.211	0.165	0.198	0.155	0.214	0.217
GSB	0.231	0.191	0.152	0.136	0.229	0.131
NB-NTM	0.188	0.223	0.248	0.245	0.127	0.125
GNB-NTM	**0.240**	0.228	0.237	0.255	0.127	0.093
HDP	0.192		0.266		0.157	
iTM-VAE	0.195		0.201		0.184	
HiTM-VAE	0.237		**0.269**		0.233	
rCRP	0.186		0.206		0.201	
hLDA	0.221		0.185		0.186	
TSNTM	0.212		0.206		0.213	
nTSNTM	0.219		0.234		0.237	
hPAM	0.213		0.229		0.223	
HSOC	0.223		0.210		–	
CluHTM	0.219		0.161		–	
Ours	0.235		0.251		**0.240**	

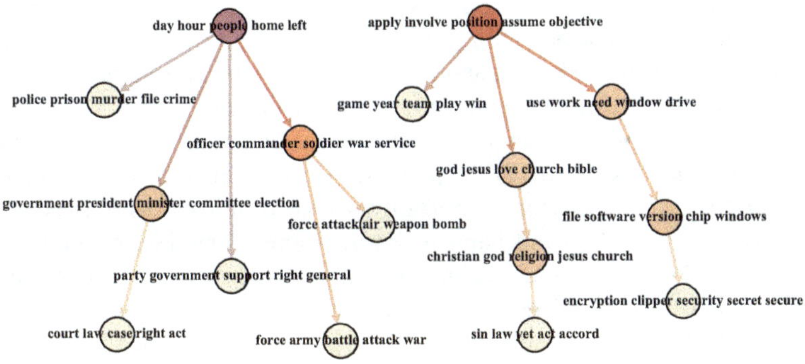

Fig. 2.24 Two examples of topic trees generated by nFNTM on Wikitext-103 (left) and 20News (right)

each topic and calculate the average NPMI scores across all models and datasets. A higher NPMI score indicates more interpretable topics. The number of topics automatically identified by our nonparametric model on the 20News, Reuters, and Wikitext-103 datasets is 61, 76, and 121, respectively.

Table 2.12 presents the NPMI scores for topics generated by various models. While our nFNTM model significantly outperforms hierarchical topic models, it falls slightly behind the top-performing flat topic model. This is due to the fact that NPMI is an average score across all topics, which may not be entirely fair to hierarchical topic models. Flat topic models treat topics independently, leading to more specific topics, whereas higher-level topics in hierarchical models tend to be more general and thus have lower NPMI scores. Figure 2.24 illustrates two

Table 2.13 Ablation evaluation on 20News

Model	#Topics	NPMI	CLNPMI
Ours	61	**0.235**	**0.152**
Ours w/o SBP	50	0.223	0.135
Ours w/o self-attention	57	0.209	0.097
Ours w/o self-attention+DAG	48	0.203	–

examples of topic trees generated by nFNTM on Wikitext-103 (left) and 20News (right), focusing on different topics such as *politics* and *war* on the left and *play*, *religion*, and *computer* on the right. Considering the left tree as an example, the NPMI score of the root topic [*day hour people home left*] is notably lower than that of the leaf topic [*force army battle attack war*], which impacts the overall result. Additionally, topics closer to the root are more general, while those closer to the leaves are more specific. Furthermore, child topics are related to their parent topics, such as *encryption* being a child of *computer* and *law* being a child of *governance*. These topic examples underscore the advantages of a forest structure compared to a tree structure, as it can not only learn reasonable parent-child relationships among topics but also generate a flexible topic hierarchy with unlimited depth and width.

Ablation Study We conducted a series of ablation experiments to ascertain the efficacy of each component within our model. Table 2.13 presents the results of these experiments when our nonparametric Factorial Nonparametric Topic Model (nFNTM) is devoid of three specific modules. Specifically, "Our Model without SBP" signifies a parametric framework employing a Gaussian distribution instead of the Stick-Breaking Process (SBP). "Our Model without Self-Attention" implies that the weighted adjacency matrix is arbitrarily assigned rather than learned through the proposed self-attention mechanism. Furthermore, "Our Model without Self-Attention and DAG" refers to reconstructing documents solely based on topic distributions derived from SBP. It is noteworthy that in the absence of a Directed Acyclic Graph (DAG), a flat topic hierarchy emerges, rendering the CLNPMI metric inapplicable. As evident from Table 2.13, the incorporation of nonparametric and self-attention modules contributes significantly to the formulation of a coherent topic hierarchy, thereby enhancing topic interpretability.

Concentration Parameter Evaluation To examine the nonparametric characteristics of our model on the 20News dataset, we undertake an analysis by varying the values of the concentration parameter $\hat{\boldsymbol{\beta}}$. Figure 2.25 demonstrates that an increase in the value of $\hat{\boldsymbol{\beta}}$ results in a corresponding surge in the number of topics detected by all nonparametric models. This finding is consistent with theoretical expectations, as a larger $\hat{\boldsymbol{\beta}}$ leads to a smoother distribution under the Stick-Breaking Process (SBP), which in turn tends to generate a greater number of topics compared to a denser distribution. Furthermore, our model exhibits a greater capacity to generate diverse topics for larger values of $\hat{\boldsymbol{\beta}}$, as these topics are more dispersed [85]. Additionally, our model outperforms other models in approximating the nonparametric properties of the Hierarchical Dirichlet Process (HDP).

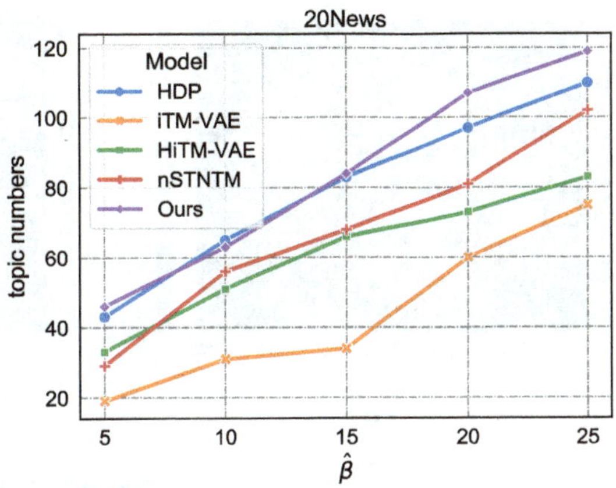

Fig. 2.25 Topic numbers derived by different nonparametric models with various values of $\hat{\beta}$

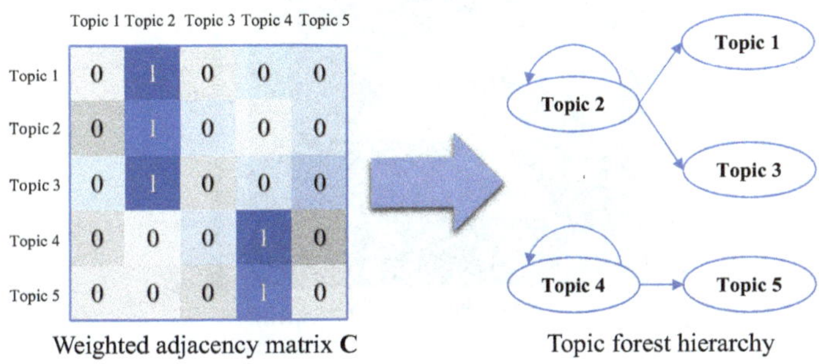

Fig. 2.26 An example of building a topic forest from a weighted adjacency matrix

Forest-Structured Topic Visualization In this section, we conduct a qualitative analysis to assess the rationality of the topic forest produced by our model. We establish parent-child topic relationships using the values in the weighted adjacency matrix **C**. Specifically, a value of $C_{i,j} \approx 1$ indicates that the child topic t_i is connected to the parent topic t_j. For illustrative purposes, we provide an example of **C** and demonstrate how to construct a topic forest from it in Fig. 2.26. This hierarchical topic structure is grounded in the assumption that a document can be decomposed into a weighted combination of multiple topics, with each topic being further decomposable into a weighted combination of subtopics [15].

Figure 2.27a visualizes the weighted adjacency matrix derived from the Reuters dataset. The white dot means that there is a correlation between two topics indexed in the row and column. Notably, apart from the dots along the diagonal, the remaining dots exhibit a discrete distribution, adhering to the constraints of a Directed Acyclic Graph (DAG). Furthermore, we observe that topic 51 is linked to numerous other topics, functioning as the parent topic for multiple others and serving as a general root topic for the entire corpus. Similarly, Fig. 2.27b, c depict the weighted adjacency matrices generated by our model on the 20News and Wikitext-103 datasets, respectively. The results indicate that, in both cases, apart from the

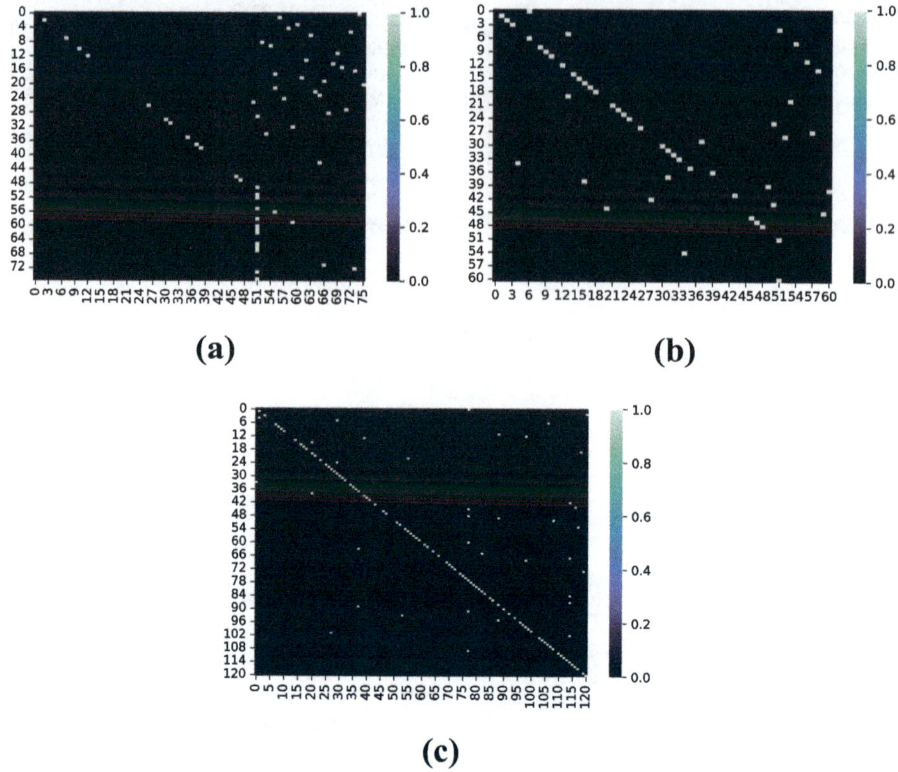

Fig. 2.27 Visualization of the weighted adjacency matrix generated from (**a**) Reuters, (**b**) 20News, and (**c**) Wikitext-103, respectively

dots on the diagonal, the remaining dots show a discrete distribution, satisfying the DAG constraints.

2.2.3 Gaussian Mixture Hierarchical Topic Model

This work proposed the NSEM-GMHTM, a deep topic model that employs a **Gaussian mixture prior** distribution to enhance the model's adaptability to sparse data. NSEM-GMHTM explicitly models hierarchical and symmetric relations between topics through dependency matrices and **nonlinear structural equations**.

2.2.3.1 Introduction

Topic modeling techniques have become ubiquitous in text analysis, serving to uncover the latent semantic structure within text corpora. Their primary objective is to identify a collection of semantically coherent topics from a set of documents. Each topic embodies a typical pattern of word co-occurrences and is often semantically interpreted as a coherent group of words representing a specific concept. While traditional models, such as Latent Dirichlet Allocation (LDA) [5] and the Embedded Topic Model (ETM) [18], have been successful in achieving this goal, they are constrained by their assumption of topic independence, which hinders their capacity to explore the intricate topic structure within a corpus. In response to this limitation, a range of hierarchical extensions have been introduced. These include hierarchical LDA (hLDA) [4], the recursive Chinese Restaurant Process (rCRP) [38], and the nested Hierarchical Dirichlet Process (nHDP) [64]. These models organize topics in hierarchical tree structures, with upper-level topics being more general or abstract, and lower-level topics being more specific. By revealing the hierarchical relationships among topics, these models provide users with a more intuitive understanding of the text data. However, they rely on approximate inference methods, such as variational inference and Gibbs sampling, which can involve complex mathematical derivations and high computational costs.

With advancements in deep neural networks and the introduction of Neural Variational Inference (NVI), there has been a surge of interest in the development of Neural Hierarchical Topic Models (NHTMs), primarily due to their efficient parameter inference and enhanced flexibility [15, 21, 35, 88]. Typically, NHTMs leverage the Variational Auto-Encoder (VAE) framework and conceptualize topic hierarchy through the relationships among neurons at various levels within the encoder or decoder, exemplified by models such as the Tree-Structured Neural Topic Model (TSNTM) [35] and its nonparametric counterpart and the nTSNTM [15].

However, a significant limitation of most NHTMs is their reliance on a single isotropic multivariate Gaussian prior distribution, which often proves inadequate in accurately approximating the posterior distributions of sparse data [87]. A more precise estimation of the posterior distribution could substantially enhance the capability of VAEs in fitting and analyzing sparse datasets. Beyond topic modeling, the utilization of the Gaussian Mixture Model (GMM) as a prior for latent variables has recently demonstrated impressive performance in domains such as image generation and bioinformatics [87, 90].

It is observed that earlier NHTMs have concentrated predominantly on the interconnections among topics across varying degrees of abstraction. A case in point, as documented by [15], involves constructing a topic tree in an ascending manner through employing a dependency matrix, wherein the overarching theme represents a generalized version of its subordinate themes. Nevertheless, it is equally important to consider how high-level topics might be affected by the configuration among lower-tier topics. In biochemical pathways or social network communities, for instance, the exchange of information between nodes within the same stratum significantly contributes to deriving higher-level abstract modules [17].

Considering the essence of topics themselves, they can be perceived as words with broad semantic meanings. According to [72], not only do hierarchical connections exist between subjects at different levels of generality, such as Chicago

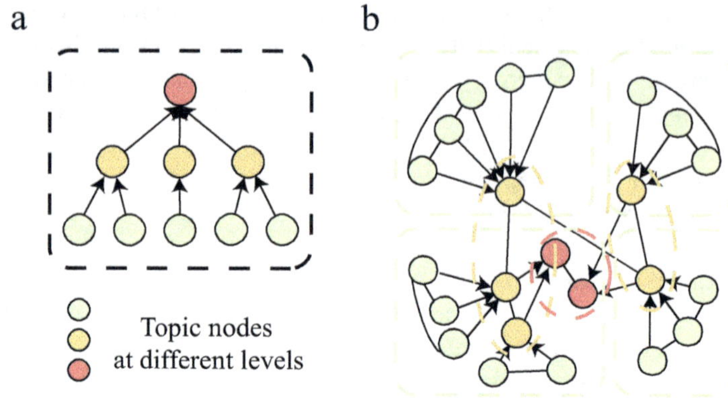

Fig. 2.28 Illustration of (**a**) topics discovered by previous hierarchical topic models and (**b**) topics found by our hierarchical topic model guided by nonlinear structural equations

and city ("ISA"), but there should also be lateral connections that link items on the same level, like cut and knife ("CapableOf") or learned and learned ("RelatedTo"). Regrettably, current NHTMs typically define topics within a tree framework, emphasizing solely the modeling of hierarchical relationships while overlooking lateral connections that could provide additional insights for researchers analyzing textual data.

Moreover, leveraging symmetrical structures within topics to enhance models' ability to grasp document semantics has received scant attention. While some studies [49, 79] have experimented with generating documents using Directed Acyclic Graph (DAG) structured topics, the resulting topic structure often lacks clarity.

To address these limitations, this study introduces the Nonlinear Structural Equation Model guided Gaussian Mixture Hierarchical Topic Model (NSEM-GMHTM), a sophisticated generative model for document analysis. As illustrated in Fig. 2.28, diverging from conventional hierarchical topic models, our approach employs a Nonlinear Structural Equation Model (NSEM) to explicitly formulate symmetric dependencies among topics. This facilitates the derivation of a more thorough and lucid topic architecture. Specifically, we adopt a Gaussian mixture distribution as the prior for latent variables, empowering the model to learn more intricate distributions, thus enhancing the capability of the Variational Auto-Encoder (VAE) in handling and analyzing sparse datasets. Experimental results demonstrate that our proposed model surpasses existing state-of-the-art benchmarks across various commonly used metrics, affirming the reasonableness of the generated topic structure.

Additionally, through ablation studies and extensive qualitative assessments, it has been shown that the application of NSEM within the neural hierarchical topic modeling framework leads to superior topic structuring. These findings further substantiate the efficacy and validity of our proposed methodology.

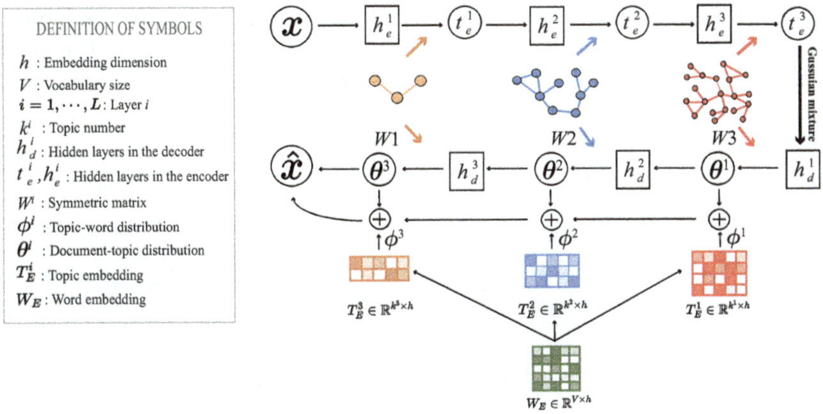

Fig. 2.29 The workflow of NSEM-GMHTM with three layers

2.2.3.2 Methodology

In this section, we introduce the NSEM-GMHTM for text analysis, which is specifically designed to uncover detailed topic architecture. The development of NSEM-GMHTM addresses two primary challenges: (i) establishing a clear construction of both symmetric and hierarchical dependencies among topics and (ii) designing expressive neural networks that enhance model adaptability and effectiveness in analyzing sparse data. Initially, we will discuss the relevant technologies involved. Following this, we detail the decoder and encoder components of the NSEM-GMHTM as illustrated in Fig. 2.29. Lastly, we elaborate on the specifics of model inference.

Gaussian Mixture VAE Variational inference has the capability to convert intractable inference issues into manageable optimization problems [80], thereby broadening the range of applicable tools to include various optimization methodologies. Nevertheless, a significant drawback of traditional variational inference is its requirement for the likelihood and prior distributions to be conjugate, which facilitates tractable optimization but simultaneously restricts the applicability of these methods.

VAE represents an innovative integration of variational inference with the adaptability and scalability provided by neural networks [41, 68]. By utilizing neural networks to produce the conditional posterior, VAE enables the optimization of the variational inference objective through stochastic gradient descent. This framework excels at learning the distribution of input data, facilitating effective combination with conventional probabilistic graphical models (e.g., LDA) and enabling rapid parameter inference [73]. However, a limitation of the standard VAE approach is its reliance on a single isotropic multivariate Gaussian prior for latent variables, which often results in underfitting when handling sparse datasets [87].

Utilizing a Gaussian Mixture Model (GMM) as the prior for latent variables has been effectively employed in unsupervised learning to produce more disentangled and interpretable latent representations. Following [19], this approach can be formulated using a joint distribution $p(\mathbf{x}, \mathbf{z}, c)$, which is factorized as

$$p(\mathbf{x}, \mathbf{z}, c) = p(\mathbf{x} \mid \mathbf{z})p(\mathbf{z} \mid c)p(c), \tag{2.43}$$

$$c \sim \text{Mult}(\pi), \tag{2.44}$$

$$\mathbf{z} \mid c \sim \prod_{k=1}^{K} \mathcal{N}\left(\mu_{c_k}, \sigma_{c_k}^2 \mathbf{I}\right)^{c_k}, \tag{2.45}$$

$$\mathbf{x} \mid \mathbf{z} \sim \mathcal{N}\left(\mu(z), \sigma^2(z)\right) \text{ or } \mathcal{B}\left(\mu(z)\right). \tag{2.46}$$

In the above, K indicates the predefined number of mixture components, $\mathbf{x}$ represents the observed data, $\mathbf{z}$ denotes the latent variable, and c is a one-hot vector sampled from the mixing probabilities π, selecting one component from the Gaussian mixture.

Nonlinear Structural Equation Model The Structural Equation Model (SEM) serves as a multivariate statistical framework aimed at analyzing the structural relationships among various random variables [8]. Initially, SEM was developed to model the covariance matrix for random variables. Over time, it has proven highly effective in modeling the relationship between observed features and latent variables, gaining widespread application in fields such as econometrics and sociology for causal inference [24, 50]. Notably, SEM can be utilized to identify conditional dependencies among random variables, making it suitable for predicting graph structures in Bayesian networks and Markov random fields [92].

Let $W \in \mathbb{R}^{D\times D}$ denote the weighted adjacency matrix for D variables (nodes), and let $X \in \mathbb{R}^{D\times h}$ represent a sample from the joint distribution of D variables over h features, with each row corresponding to one variable. The linear SEM model is given by

$$X = W^T X + Z. \tag{2.47}$$

In the above, $Z \in \mathbb{R}^{D\times h}$ represents a noise matrix that follows a Gaussian distribution. By integrating traditional linear SEM with deep learning techniques capable of capturing complex nonlinear mappings, a nonlinear variant of SEM (termed NSEM) was proposed by [92]. It can be formulated as

$$X = f_1\left(\left(\mathbf{I} - W^T\right)^{-1} Z\right), \tag{2.48}$$

$$Z = \left(\mathbf{I} - W^T\right) f_2(X). \tag{2.49}$$

In these equations, **I** signifies the identity matrix, while $f_1(\cdot)$ and $f_2(\cdot)$ denote multilayer neural networks. Extending NSEM into bioinformatics, [71] successfully predicted regulatory relationships between genes, demonstrating that this approach can aid models in capturing symmetric dependencies between topics.

Modeling Process Motivated by preceding research, we incorporate a Gaussian Mixture Model (GMM) as the prior for latent variables. This enables the network to learn more intricate distributions and enhances the model's capability to fit and analyze sparse datasets. Furthermore, aiming to uncover a richer topic architecture from a corpus of documents, NSEM-GMHTM builds upon the Nonlinear Structural Equation Model (NSEM) introduced by [92]. Our extension specifically targets capturing symmetric dependencies among topics at identical levels. The specifics of our proposed model are detailed below.

Document Encoder Given a set of documents, each document is processed into a bag-of-words (BoW) vector $\boldsymbol{x}_{\text{bow}} \in \mathbb{R}^V$, where V represents the vocabulary size. According to the definition by [19], the Gaussian mixture encoder network can be described as follows:

$$h_e^1 = f_1(\boldsymbol{x}_{\text{bow}}), \tag{2.50}$$

$$t_e^i = \left(\left(\mathbf{I} - |W^i|^T\right)(h_e^i)^T\right)^T, \tag{2.51}$$

$$h_e^{i+1} = \tanh\left(f_2\left(t_e^i\right)\right), \tag{2.52}$$

$$c = \text{Gumbel Softmax}(t_e^L), \tag{2.53}$$

$$t_d^1 = \text{Reparameterize}\left(t_e^L, c\right). \tag{2.54}$$

In these equations, the Gumbel Softmax layer outputs a K-dimensional label, with its i_{th} dimension representing the probability that the input vector belongs to the i_{th} Gaussian mixture component. During training, these probabilities are progressively concentrated on a single component [36]. Following [19], we set the number of mixture components K to 10. For each topic layer, both $(\mathbf{I} - |W|^T)$ and $(\mathbf{I} - |W|^T)^{-1}$ are trained in the encoder and decoder to capture symmetric relationships among topics, enhancing the model's ability to understand the implicit semantic structure of the corpus. It should be noted that h_e^i and t_e^i represent hidden features before and after incorporating the symmetric relation, respectively.

Document Decoder Considering the generative model of NSEM-GMHTM with L layers, from bottom to top, the document decoder can be formulated as follows:

$$h_d^i = \left(\left(\mathbf{I} - |W^i|^T\right)^{-1}(t_d^i)^T\right)^T, \tag{2.55}$$

$$t_d^{i+1} = \tanh\left(h_d^i|M^i|\right), \tag{2.56}$$

$$\boldsymbol{\theta}^i = \text{softmax}\left(h_d^i\right), \tag{2.57}$$

$$\boldsymbol{\phi}^i = \text{softmax}\left(T_E^i \times W_E\right), \tag{2.58}$$

$$\hat{\boldsymbol{x}} = \sum_{i=1}^{L} \hat{\boldsymbol{x}}^i = \sum_{i=1}^{L} \boldsymbol{\theta}^i \boldsymbol{\phi}^i. \tag{2.59}$$

In the above, $W^i \in \mathbb{R}^{k^i \times k^i}$ is a symmetric matrix, $M^i \in \mathbb{R}^{k^i \times k^{i+1}}$ is a dependency matrix designed to capture hierarchical relationships among topics at various levels, and k^i indicates the number of topics at layer i. It is important to note that the weights of W^i and M^i are constrained to be nonnegative to preserve interpretability regarding the directionality of topic structure. The topic-word distribution $\boldsymbol{\phi}^i$ is computed using Eq. (2.58), which involves topic embeddings T_E^i and word embeddings W_E. Subsequently, we reconstruct the document $\hat{\boldsymbol{x}}^i$ by combining the document-topic distribution $\boldsymbol{\theta}^i$ with the topic-word distribution $\boldsymbol{\phi}^i$. To ensure each layer contributes independently, we configure the decoder to reconstruct each layer back into an $\hat{\boldsymbol{x}}^i$.

Gaussian Mixture Prior Traditional topic models frequently depend on a single isotropic multivariate Gaussian prior, which might be insufficient for capturing the complexities of sparse datasets. To overcome this limitation, we introduce a Gaussian Mixture Model (GMM) as the prior over latent variables. This method allows the model to learn more sophisticated distributions and improves its capability to fit and analyze sparse textual data. The GMM is characterized by a mixture of several Gaussian distributions, each with its own mean and covariance matrix. The joint distribution $p(\boldsymbol{x}, z, \boldsymbol{c})$ can be factorized as

$$p(\boldsymbol{x}, z, \boldsymbol{c}) = p(\boldsymbol{x}|z)p(z|\boldsymbol{c})p(\boldsymbol{c}), \tag{2.60}$$

where $\boldsymbol{x}$ represents the input variable, z denotes the latent variable, and $\boldsymbol{c}$ is a one-hot vector sampled from the mixing probability $\boldsymbol{\pi}$, selecting one component from the Gaussian mixture.

The specific forms of these distributions are given by

$$\boldsymbol{c} \sim \text{Mult}(\boldsymbol{\pi}), \tag{2.61}$$

$$z|\boldsymbol{c} \sim \sum_{k=1}^{K} \mathcal{N}(\boldsymbol{\mu}_{ck}, \boldsymbol{\sigma}_{ck}^2 \mathbf{I}), \tag{2.62}$$

$$\boldsymbol{x}|z \sim \mathcal{N}(\boldsymbol{\mu}(z), \boldsymbol{\sigma}^2(z)) \text{ or } \mathcal{B}(\boldsymbol{\mu}(z)). \tag{2.63}$$

In the above, K indicates the predefined number of components in the mixture, $\boldsymbol{\pi}$ represents the mixing probabilities, $\boldsymbol{\mu}_{ck}$ and $\boldsymbol{\sigma}_{ck}^2$ denote the mean and variance of the kth Gaussian component for class c, and $\mathbf{I}$ is the identity matrix of appropriate dimension.

Nonlinear Structural Equation Traditional topic models typically assume a hierarchical structure resembling a tree, thereby neglecting symmetric dependencies among topics at the same level. To address this limitation and capture both hierarchical and symmetric relationships, we employ Nonlinear Structural Equations. This method enables us to explicitly model inter-topic dependencies and gain deeper insights into the semantic structure of textual data.

The Nonlinear Structural Equation Model (NSEM) is utilized for modeling the relationship between observed features and latent variables. The fundamental form of NSEM can be expressed as

$$\boldsymbol{x} = f_1(\mathbf{W}^T \boldsymbol{x} + z), \tag{2.64}$$

$$z = f_2(\boldsymbol{x}), \tag{2.65}$$

where $\boldsymbol{x}$ represents the matrix of observed variables, $\mathbf{W}$ denotes the weighted adjacency matrix that captures the relationships among variables, and z is a noise matrix following a Gaussian distribution. Functions f_1 and f_2 are multilayer neural networks designed to model nonlinear mappings.

By integrating GMM with NSEM, our proposed model, NSEM-GMHTM, effectively captures the intricate relationships between topics. This integration enhances the interpretability and coherence of the topic structure while improving the model's capability to fit sparse data. Consequently, it offers a more thorough understanding of the semantic relationships within the corpus.

2.2.3.3 Experimental Results

Datasets Our experimental evaluations are performed on three commonly utilized text datasets, each differing in size. These datasets include 20News [54], NIPS [74], and Wikitext-103 [61]. Prior to analysis, all datasets underwent preprocessing steps, which involved the removal of stop words and elimination of low-frequency terms. The detailed statistics for each dataset are summarized in Table 2.14.

Baselines and Parameter Settings For hierarchical topic models, we utilize TSNTM [35], CluHTM [79], SawETM [21], HyperMiner [88], and nTSNTM [15]

Table 2.14 The statistics of datasets

Dataset	#Docs (Train)	#Docs (Test)	Vocabulary size
20News	11,314	7531	3997
NIPS	1350	149	3531
Wikitext-103	28,472	120	20,000

as our baseline models. For all these models, the maximum depth of the topic hierarchy is set to 3, following [35]. To ensure a fair comparison among hierarchical topic models, we configure each model with an identical number of topics.

For nonparametric or flat topic models, we adopt HDP [75], ETM [18], NB-NTM and GNB-NTM [85], and iTM-VAE and HiTM-VAE [62] as baselines. HDP is a classical nonparametric topic model that allows for a potentially infinite number of topics. ETM combines LDA [5] with word embeddings, assuming that topics and words coexist in the same embedding space, thus learning interpretable word and topic embeddings. The iTM-VAE and HiTM-VAE models extend the methodology from [60] by integrating nonparametric processes into the NVI framework to extract an infinite number of potential topics.

To facilitate a fair comparison between parametric and nonparametric topic models, we follow [15] to set the number of topics to 50 and 200 for all flat parametric models. For nonparametric models (HDP, iTM-VAE and HiTM-VAE, CluHTM, and nTSNTM), we use the optimal hyperparameters reported in their respective original papers. For parametric hierarchical topic models (SawETM, HyperMiner, and NSEM-GMHTM), the number of topics at different layers is set to 128, 32, and 8. It should be noted that for all metrics except topic specialization [38], we compute the average score for the top-5, -10, and -15 words.

Evaluation on Topic Interpretability In this part, we use the widely adopted NPMI [9, 15, 35, 79] to evaluate topic interpretability. As mentioned by [43], NPMI is a measurement of topic coherence that is closely consistent with the ranking of topic interpretability by human annotators. As shown in Table 2.15, the proposed model performs significantly better than previous NHTMs on all datasets, achieving a better NPMI by a margin of 8.9% on 20News, 26.7% on NIPS, and 7.6% on Wikitext-103, where percentage improvements are determined over the second best NHTMs. Compared to SawETM, our NSEM-GMHTM's NPMI is

Table 2.15 The average NPMI scores between top-5, -10, and -15 words in each topic

Dataset	20News		NIPS		Wikitext-103	
Model	50	200	50	200	50	200
ETM	0.263	0.248	0.098	0.068	0.214	0.217
NB-NTM	0.265	0.281	0.107	0.103	0.127	0.125
GNB-NTM	0.292	0.278	0.101	0.126	0.127	0.093
HDP	0.273		0.131		0.157	
iTM-VAE	0.278		0.098		0.184	
HiTM-VAE	0.294		0.135		0.233	
SawETM	0.264		0.133		0.154	
nTSNTM	0.262		0.101		0.169	
TSNTM	0.282		0.116		0.237	
HyperMiner	0.263		0.135		0.225	
CluHTM	0.219		0.122		–	
NSEM-GMHTM	**0.307**		**0.147**		**0.255**	

on average 30.8% higher across the three datasets, presumably because SawETM only constructs topic hierarchies, which are not entirely accurate, whereas NSEM-GMHTM novelly models symmetric relationships of topics and is therefore able to capture the structural properties between topics at the same level. In addition, our method shows competitive performance compared to the best flat baselines. In particular, the NPMI of NSEM-GMHTM is improved by 8.9% compared to HiTM-VAE on the NIPS dataset. We do not report the results of CluHTM on Wikitext-103 since it failed to achieve convergence in 48 hours.

Topic Structure Analysis In this section, we use the evaluation metrics proposed by prior works, including topic specialization [38], cross-level normalized pointwise mutual information (CLNPMI) [15], topic uniqueness (TU) [61], and overlap rate (OR) [15] to comprehensively assess the topic hierarchy generated by NSEM-GMHTM from different perspectives and to compare it with state-of-the-art approaches. Key words of topics are ranked from the topic-word matrix $\boldsymbol{\phi}$ [4].

Semantic Rationality of Topic Hierarchy In the real world, higher-level topics serve as generalized representations of their lower-level counterparts. For example, *basketball* and *football* can be categorized under the broader topic of *sports*. Thus, topics at higher levels should exhibit more general semantics, whereas those closer to the bottom should be more specific. Topic specialization [38] measures this characteristic by computing the cosine distance between each topic's word distribution and that of the entire corpus. A higher score indicates greater specialization. Therefore, we use topic specialization as a metric for evaluating the semantic rationality of the topic hierarchy. Figure 2.30 illustrates the topic specialization scores across all hierarchical topic models at each level. The results demonstrate that NSEM-GMHTM achieves a coherent pattern of topic specialization across different datasets, with scores decreasing as the depth of the hierarchy increases. Conversely, CluHTM shows nearly constant topic specialization scores close to 1 at all levels on the 20News and NIPS datasets, suggesting an unreasonable topic hierarchy.

Furthermore, when a model struggles to capture the complex underlying structure within a corpus, it may suffer from mode collapse, leading to overly similar topics. To measure the semantic redundancy in topic hierarchies, we employ the widely used metric of topic uniqueness (TU) [61], calculated as follows:

$$\mathrm{TU}(k) = \frac{1}{NK}\sum_{k=1}^{K}\sum_{n=1}^{N}\frac{1}{\mathrm{cnt}(n,k)}, \tag{2.66}$$

where K is the number of topics and $\mathrm{cnt}(n, k)$ denotes the frequency of the nth top word in the kth topic appearing among the top N words across all topics. Results presented in Table 2.16 indicate that our model significantly outperforms baseline methods. These findings suggest that NSEM-GMHTM generates topics with clear hierarchy and minimal redundancy, highlighting its effectiveness in ensuring semantic rationality within topic hierarchies.

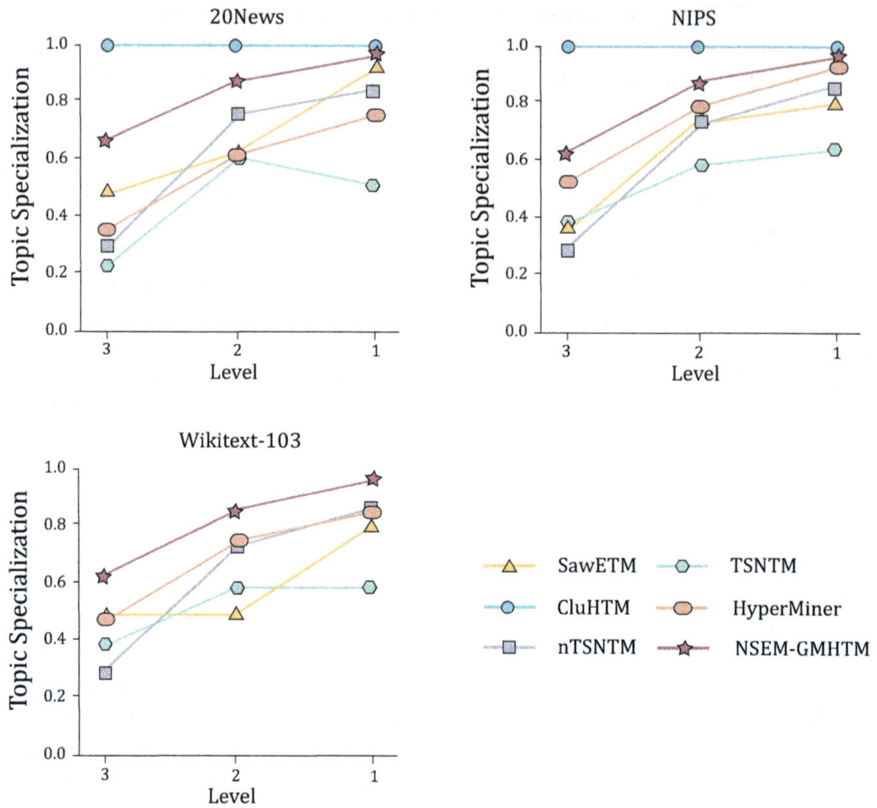

Fig. 2.30 Topic specialization of different hierarchical topic models at each level

Structural Rationality of Topic Hierarchy As highlighted by a previous work [79], a well-structured topic hierarchy should ensure that child topics are coherent with their corresponding parent topics, yet not overly similar to them. To evaluate this coherence and redundancy between parent and child topics, we utilize Cross-Level Normalized Pointwise Mutual Information (CLNPMI) and Overlap Rate (OR), respectively.

CLNPMI, introduced by [15], measures the average NPMI value between every parent topic and its children using the formula:

$$\mathrm{CLNPMI}(W_p, W_c) = \sum_{w_i \in W'_p} \sum_{w_j \in W'_c} \frac{\mathrm{NPMI}(w_i, w_j)}{|W'_p||W'_c|}, \tag{2.67}$$

where $W'_p = W_p - W_c$ and $W'_c = W_c - W_p$. Here, W_p and W_c represent the top-N words of a parent topic and a child topic, respectively.

Overlap rate (OR) quantifies the average repetition ratio of the top-N words between parent topics and their children, defined as

Table 2.16 The CLNPMI, TU, and OR scores of all hierarchical topic models, where "–" indicates that the results could not be obtained within 48 hours

Dataset	Metric	SawETM	CluHTM	TSNTM	nTSNTM	HyperMiner	NSEM-GMHTM
Wikitext-103	CLNPMI ↑	0.060	–	0.086	**0.113**	0.079	0.090
	TU ↑	0.221	–	0.615	0.730	0.520	**0.797**
	OR ↓	0.064	–	0.078	0.080	0.162	**0.017**
20News	CLNPMI ↑	0.138	0.123	0.109	0.144	0.143	**0.146**
	TU ↑	0.716	0.577	0.430	0.683	0.388	**0.811**
	OR ↓	0.064	0.332	0.052	0.030	0.143	**0.011**
NIPS	CLNPMI ↑	0.034	0.098	0.113	0.022	**0.048**	0.028
	TU ↑	0.431	0.285	0.116	0.373	0.662	**0.719**
	OR ↓	0.071	0.447	0.078	0.063	0.135	**0.025**

$$OR = \frac{|W_p \cap W_c|}{N}. \tag{2.68}$$

Following a previous research [21], we consider the two most relevant lower-level topics for each upper-level topic as parent-child pairs.

Table 2.16 presents the performance of different models across multiple datasets. The results demonstrate that our model ensures the highest diversity while maintaining strong coherence between parent and child topics, thereby proving the structural rationality of the topic hierarchy.

Visualization of the Embedding Space The top-5 words from eight topics generated by NSEM-GMHTM on the 20News dataset are visualized in Fig. 2.31 using UMAP visualization [52]. This visualization demonstrates that the topics are highly interpretable within the word embedding space, with each topic being proximate to semantically related terms.

Notably, words belonging to the same topic cluster closely together, whereas those from different topics remain distinctly separated. Moreover, related topics also exhibit proximity in the embedding space. For instance, Topic: 1_94 and Topic: 1_126 are positioned close to each other, indicating their semantic similarity. This spatial arrangement supports the model's ability to capture meaningful relationships among topics and their constituent words.

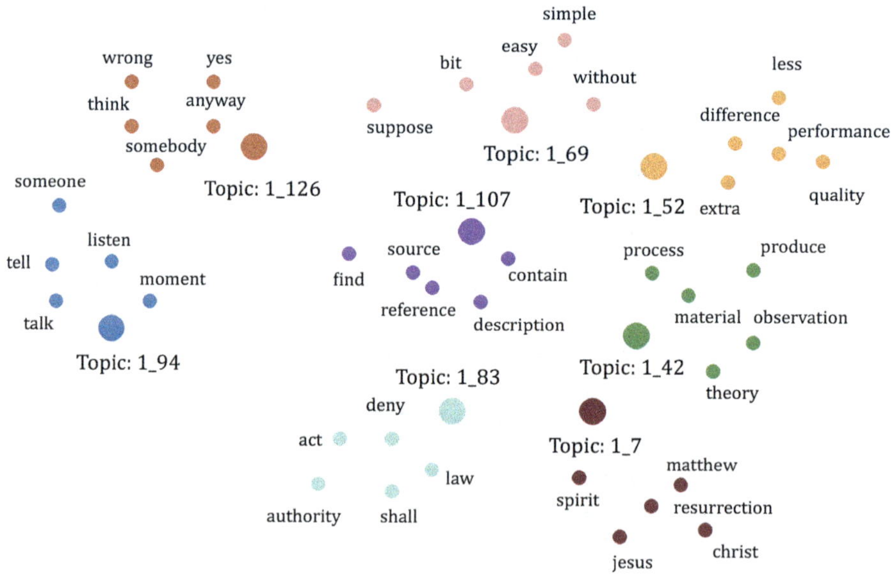

Fig. 2.31 Visualization of word and topic embeddings, where Topic: l_i denotes the i_{th} topic at the l_{th} layer

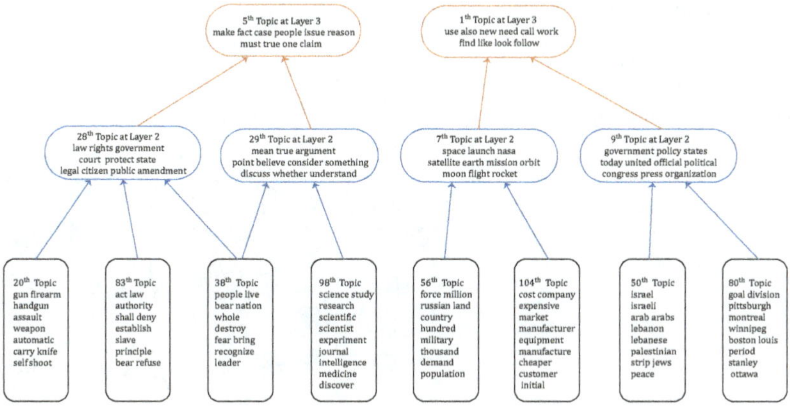

Fig. 2.32 An example of hierarchical topics learned from 20News by NSEM-GMHTM

Hierarchical Structure of Topics To visually demonstrate the capability of our model in generating hierarchical topic structures (i.e., relationships between topics at different levels), we present a visualization of several topics extracted by NSEM-GMHTM from the 20News dataset. As illustrated in Fig. 2.32, each rectangle represents a topic along with its top-10 words, and arrows indicate the relationship from subtopics to their most closely related parent topics.

Consistent with the concept of topic specialization, topics closer to the root are more general, whereas those nearer to the leaves are more specific. Furthermore, child topics are semantically related to their parent topics. For example, *boston* is a child topic under *states*, and *authority* falls under *law*. These results show that the semantic meaning of each topic, and the connections between adjacent layers are highly interpretable, indicating that our method can effectively learn a reasonable topic hierarchy.

Symmetric Structure of Topics Besides its proficiency in topic hierarchy learning, another notable feature of NSEM-GMHTM is its capability to uncover an interpretable symmetric structure among topics. In this section, we conduct a topic symmetric relation discovery experiment on the 20News dataset. We query the top-ranked same-level topic associations, with some examples presented in Table 3.10. The results indicate that our model can effectively capture symmetric dependencies between topics, such as *nsa* for Topic: 1_76 and *secure* for Topic: 1_99, as well as *israel* and *nazi* for the seventh-ranked association.

Furthermore, we extract topic symmetric dependencies from the first layer and construct a topic-topic network by selecting 100 topic associations with the greatest weight as edges. To facilitate the analysis of this network, we utilize Gephi[6] to visualize the topics and identify communities using a community detection

[6] https://gephi.org/

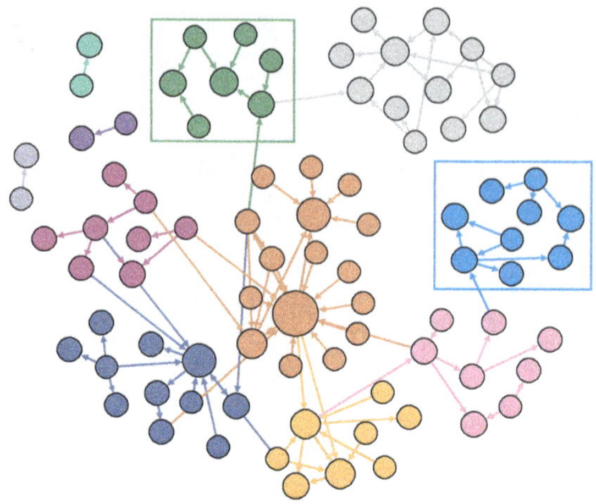

Fig. 2.33 First-layer topic network structure learned from 20News dataset using NSEM-GMHTM with communities represented by distinct colors

Table 2.17 Top-5 words of topics in the green box of Fig. 2.33

Label					
Topic: 1_82	cost	market	costs	cheaper	expensive
Topic: 1_70	armenian	armenians	armenia	azerbaijan	genocide
Topic: 1_100	jews	jewish	greek	adam	jew
Topic: 1_126	surrender	banks	pitt	gordon	intellect
Topic: 1_106	nazi	muslim	german	genocide	nazis
Topic: 1_25	religion	atheism	morality	atheists	religious
Topic: 1_79	israel	israeli	arab	palestinian	lebanon

Table 2.18 Top-5 words of topics in the blue box of Fig. 2.33

Label					
Topic: 1_76	clipper	des	nsa	escrow	encrypted
Topic: 1_99	encryption	privacy	secure	rsa	cryptography
Topic: 1_64	low	high	rate	higher	rates
Topic: 1_91	state	rights	constitution	political	civil
Topic: 1_32	anti	population	armed	murder	crime
Topic: 1_53	crime	fbi	batf	waco	defense
Topic: 1_43	gun	guns	firearms	weapon	handgun
Topic: 1_40	crime	fbi	batf	waco	defense
Topic: 1_125	key	keys	blocks	pgp	scheme

algorithm [7]. As illustrated in Fig. 2.33, topics with symmetric associations tend to form clusters, which exhibit tighter semantics within each cluster (see Table 2.17 and Table 2.18). This suggests that exploring symmetric associations between topics could be beneficial for further mining the semantic structure of a text corpus.

Ablation Study For analyzing the effect of each component of our model, we conduct ablation studies in three cases: (1) without replacing the Gaussian prior

Table 2.19 Results of ablation evaluation on all datasets

Datasets	Model	NPMI ↑	TU ↑	OR ↓	CLNPMI ↑
Wikitext-103	Ours	0.255	**0.791**	**0.017**	0.090
	Ours w/o GMM	0.255	0.641	0.021	0.092
	Ours w/o PWE	0.252	0.787	0.025	0.011
	Ours w/o NSEM	**0.261**	0.641	0.045	**0.147**
20News	Ours	**0.307**	**0.811**	**0.011**	0.146
	Ours w/o GMM	0.271	0.436	0.016	0.131
	Ours w/o PWE	0.277	0.698	0.019	0.127
	Ours w/o NSEM	0.284	0.807	0.038	**0.171**
NIPS	Ours	**0.147**	**0.719**	**0.028**	0.025
	Ours w/o GMM	0.129	0.642	0.031	0.025
	Ours w/o PWE	0.126	0.681	0.037	0.031
	Ours w/o NSEM	0.141	0.689	0.042	**0.057**

Table 2.20 Speed and number of parameters for NHTMs on the 20News dataset

Metric	SawETM	TSNTM	nTSNTM	HyperMiner	NSEM-GMHTM
Speed	5.2S	11.3S	38.6S	4.4S	3.8S
#Params	1.9M	1.3M	0.5M	2.2M	1.5M

distribution with a Gaussian mixture distribution (w/o GMM), (2) without using pretrained word embeddings (w/o PWE), and (3) without introducing NSEM to capture symmetric topic relations (w/o NSEM). Table 2.19 summarizes all metrics on the three datasets for these three cases with NSEM-GMHTM.

The results suggest that by incorporating pretrained word embeddings (PWE) and a Gaussian mixture model (GMM), the model can more effectively capture the underlying topic hierarchy and achieve improvements in topic interpretability. Additionally, the introduction of NSEM enhances the semantic coherence among topics but slightly reduces their uniqueness. Notably, NSEM has a significant impact on Topic Uniqueness (TU), Normalized Pointwise Mutual Information (NPMI), and Overlap Rate (OR), indicating that exploring symmetrical relationships between topics can help generate a more rational topic structure and improve the overall interpretability of the model.

In summary, all components of the NSEM-GMHTM method are both reasonable and effective, contributing to its superior performance in topic modeling.

Analysis of Model Complexity Here, we compare the complexity of our model and all benchmarks of NHTMs. Specifically, we average the cost of ten training epochs for each model on 20News to record the running time. In addition, the number of parameters for models is recorded, as shown in Table 2.20. It is worth noting that, although CluHTM is excluded due to its unique training strategy, it is clear from Table 2.16 that its running time is far greater than that of the other NHTMs. We can find that NSEM-GMHTM achieves competitive performance, demonstrating that explicitly modeling hierarchical and symmetric dependencies does not significantly increase the complexity of the model and further demonstrating the scalability of our model.

2.2.4 Forest-Structured Adversarial Neural Topic Model

Topic modeling can effectively discover the semantics structure of documents. However, most methods model the corpus as an independent topic structure without considering the possible hierarchical dependence between topics. In the aspect of hierarchical topic modeling, the existing neural topic models use either the double recurrent neural network or a dependency matrix to explicitly model the topic tree. However, these models need to preset the structure of the tree manually, rendering poor model flexibility. To overcome the above limitation, this study proposes a new hierarchical neural topic model named FANTM. Particularly, we model the topic hierarchy as a forest by introducing the self-attention mechanism to construct the adjacency matrix of the relationship between topics. Experiments on three real-world corpus validate the effectiveness of the proposed model.

2.2.4.1 Introduction

As a well-known text mining tool, topic modeling can effectively discover the latent semantic structure of documents. Traditional topic models such as Latent Dirichlet Allocation (LDA) [5] use Gibbs sampling [26] to estimate parameters, while neural topic models based on Variational Auto-Encoder (VAE) [48, 53] or Generative Adversarial Network (GAN) [22, 32, 83] use gradient descent for training.

Although they are trained by gradient descent, these flat topic models [5] assume that the topic distributions in documents are unrelated, and they do not model the relationships between topics, resulting in their inability of mining the corpus for potential topic hierarchies. Note that in real-world datasets, correlations between topics are very common and important [4]. A reasonable topic hierarchy helps to build a multilevel catalog of web pages [56], extract aspect-sentiment hierarchy of comments [39], and discover research topic hierarchy in academic knowledge base [64], showing high empirical value.

To compensate for the shortcomings of flat topic models, we adopt the hierarchical topic model in this work. The hierarchical topic model generates topics that are no longer independent but organized in a hierarchical structure. Existing hierarchical topic models based on probabilistic graphical models mainly obtain topic trees [4] by employing nested dining room processes (nCRP) or model topic relationships as directed acyclic graphs (DAGs) [55]. Although these methods have achieved great success in many fields, their application on large-scale datasets is challenging due to the high computational cost.

Taking advantage of the efficiency of neural networks, [35] proposed TSNTM to parameterize the hierarchical topic distribution by introducing a double recurrent

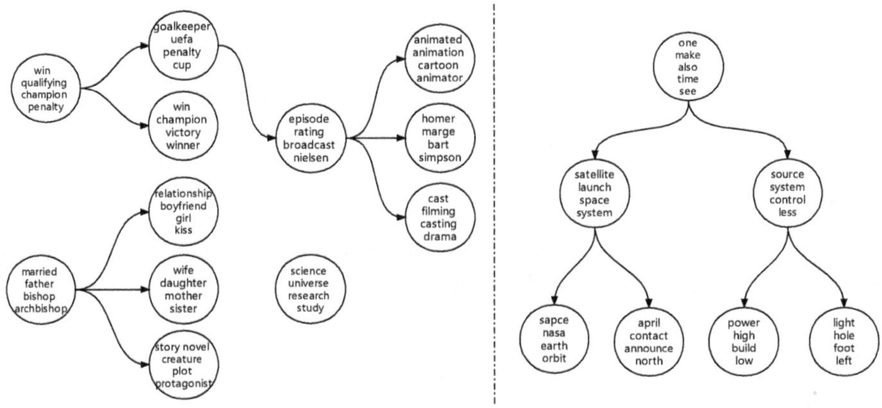

Fig. 2.34 An example of a forest-structured topic hierarchy versus a tree-like topic hierarchy

neural network (DRNN) [1]. Chen et al. [15] proposed nTSNTM, which combines a dependency matrix capturing the relationships between topic layers and a VAE. However, both TSNTM and nTSNTM require a predefined tree structure and can only generate a topic tree with a single node, which may result in a generated topic tree not conforming to the inherent topic hierarchy in the corpus.

We argue that a reasonable topic hierarchy is more likely to cover multiple topic trees (i.e., a forest-structured topic hierarchy), as shown in the left of Fig. 2.34. Nevertheless, it is difficult for the existing hierarchical topic models to construct such a topic hierarchy with a flexible structure.

In this study, we propose a **F**orest-structured **A**dversarial **N**eural **T**opic **M**odel (FANTM) to generate a topic hierarchy with flexible topic trees which can better discover the latent semantic structure of documents. Following [83, 84], we adopt GAN-based neural topic models to construct topic forests due to their powerful generative ability. We capture the relationship between topics by introducing a self-attention mechanism [78] and construct an adjacency matrix corresponding to the topic relationship. Then the corresponding topic forest is constructed through applying the Breadth-First Search (BFS) algorithm on the adjacency matrix. Our main contributions are summarized as follows:

- We are the first to propose a GAN-based hierarchical neural topic model. We introduce spectral normalization and Wasserstein distance to ensure stable training.
- We employ the self-attention mechanism to capture the hierarchical relationship between topics via an adjacency matrix and model the topic hierarchy as a flexible forest structure, which can better fit to the real topic structure of the corpus.
- We evaluate our FANTM on three datasets, and the experimental results validate the effectiveness of FANTM. We also provide in-depth discussions and visualizations for a better understanding of the generated topic forest.

2.2.4.2 Methodology

By introducing the topic self-attention mechanism to construct the adjacency matrix corresponding to the forest structure, we propose FANTM based on the GAN to model the topic hierarchy via bidirectional training. FANTM consists of an Encoder, a Generator, and a Discriminator, where the generator contains a topic self-attention mechanism module.

Encoder The encoder aims to reduce the dimension of the document distribution and obtain the topic distribution. Note that GAN does not perform well on discrete data since the backpropagation of gradients becomes difficult. Thus, TF-IDF is employed to tokenize the documents as the input for the GAN-based neural topic model [83]. For a document from the corpus, it can be tokenized as a TF-IDF vector $\boldsymbol{d}$.

Then the encoder maps $\boldsymbol{d}$ to topic distribution $\tilde{\boldsymbol{\pi}}$ through two fully connected layers.

Generator The generator is designed to reconstruct the original document from the topic distribution. We follow the previous work [83, 84] to introduce the Dirichlet distribution $\mathrm{Dir}(\boldsymbol{\alpha})$ as the prior distribution and sample the topic distribution $\boldsymbol{\pi}$ as the input of the generator as follows:

$$\boldsymbol{\pi} \sim \mathrm{Dir}(\boldsymbol{\alpha})$$
$$\mathrm{Dir}(\boldsymbol{\alpha}) \triangleq \frac{1}{\Delta(\boldsymbol{\alpha})} \prod_{s=1}^{S} \pi_s^{\alpha_s - 1}, \tag{2.69}$$

where $\boldsymbol{\alpha}$ is a preset hyperparameter of the Dirichlet distribution, S is the number of topics, $\boldsymbol{\pi}_s$ represents the proportion of the topic s in the corresponding topic distribution occupied, where $\sum_1^S \boldsymbol{\pi}_s = 1$, and $\Delta(\boldsymbol{\alpha})$ satisfies $\Delta(\boldsymbol{\alpha}) = \frac{\prod_{k=1}^{K} \Gamma(\alpha_k)}{\Gamma\left(\sum_{k=1}^{K} \alpha_k\right)}$.

To exploit the self-attention mechanism to capture the relationship between topics, we decompose the topic-word matrix $\boldsymbol{\Phi}$ into a topic embedding matrix $\boldsymbol{T_E}$ and a word embedding matrix $\boldsymbol{W_E}$ and introduce topic adjacency matrix $\boldsymbol{C}$ to construct a topic hierarchy. Finally, the generator module outputs the reconstructed document $\tilde{\boldsymbol{d}}$ corresponding to the topic distribution $\boldsymbol{\pi}$.

Discriminator The discriminator module of FANTM needs to combine the document-topic distribution and the topic-word distribution to make judgments [20], which receives fake samples $[\boldsymbol{\pi}, \tilde{\boldsymbol{d}}]$ of dimension $V+S$ or real samples $[\tilde{\boldsymbol{\pi}}, \boldsymbol{d}]$, where V represents the vocabulary size and S represents the number of topics. The input and output of the generator and the encoder are spliced together and then handed over to the discriminator to judge if it is a real sample, so that the discriminator can make decisions based on the two distributions.

Proof of Spectral Normalization in Generative Adversarial Network For any discriminator D, if there is a constant value k such that D satisfies the following inequality given any two samples $x, y \in R$, then we consider this discriminator to be k-Lipschitz continuous.

$$||D(x) - D(y)|| \leq k\|x - y\|, \tag{2.70}$$

where $\|\cdot\|$ is L2 norm. If k takes the minimum value, then k is called Lipschitz constant. Lipschitz condition limits the violent degree of function change, that is, the gradient of the function. For example, in one-dimensional space, it is easy to find that $y = cos(x)$ is 1-Lipschitz continuous, meaning that its maximum slope is 1.

Let us first prove that as long as the spectral norm of each layer W is limited to 1, the final f_Θ function will satisfy the 1-Lipschitz restriction.

$$\begin{aligned} \|Ax\| \leq L\|x\| &\Rightarrow \langle Ax, Ax\rangle \leq k^2\langle x, x\rangle \\ &\Rightarrow \langle (A^T A - k^2)x, x\rangle \leq 0. \end{aligned} \tag{2.71}$$

Let v_i denote the basis of vector and λ_i denote the eigenvalues of W, and the constraints of k are shown as follows:

$$\begin{aligned} \left\langle \left(A^T A - k^2\right) x, x\right\rangle &= \langle \left(A^T A - k^2\right) \sum_{i=1}^{n} x_i \cdot v_i, \sum_{j=1}^{n} x_j \cdot v_j\rangle \\ &= \sum_{i=1}^{n} \left(\lambda_i - k^2\right) x_i^2 \leq 0 \\ &\Rightarrow \sum_{i=1}^{n} \left(k^2 - \lambda_i\right) x_i^2 \geq 0. \end{aligned} \tag{2.72}$$

Then, we can obtain the following inequality:

$$k \geq \sqrt{max\{\lambda_i\}}. \tag{2.73}$$

Therefore, to ensure that matrix A meets 1-Lipschitz continuity, we only need to divide all elements of matrix A by $\sqrt{max\{\lambda_i\}}$ at the same time. According to the definition of spectral norm, $\sigma(W)$ is equal to $\sqrt{max\{\lambda_i\}}$, and the above requirement can be satisfied.

Topic Attention Mechanism Inspired by the word embedding mechanism [65], we introduce the topic embedding $\mathbf{T_E} \in \mathbb{R}^{S \times d_t}$ as the representation of topics, where S is the number of topics and d_t is the dimension of topic embeddings. Considering that the relevance of each topic should be learned independently, each topic embedding is projected to a different subspace, which can be formulated as follows:

$$\begin{aligned} \mathbf{Q} &= \mathbf{T_E W_Q}, \\ \mathbf{K} &= \mathbf{T_E W_K}, \end{aligned} \tag{2.74}$$

where $\mathbf{W_Q} \in \mathbb{R}^{d_t \times d_r}$ and $\mathbf{W_K} \in \mathbb{R}^{d_t \times d_r}$ are trainable parameters, and d_r is the dimension of each topic embedding subspace. $\mathbf{Q} \in \mathbb{R}^{S \times d_r}$ and $\mathbf{K} \in \mathbb{R}^{S \times d_r}$ represent the Query and Key matrices in the self-attention mechanism.

Based on $\mathbf{Q}$ and $\mathbf{K}$, we generate a weighted adjacency matrix $\mathbf{C} \in \mathbb{R}^{S \times S}$, representing the mutual relationship between topics by $\hat{C} = \mathbf{QK}^T$ and Eq. (2.75). The element $C_{i,j}$ represents the parent-child relevancy between topics t_i and t_j, which is calculated as follows:

$$C_{i,j} = \frac{exp(\frac{\hat{C}_{i,j}}{\tau})}{\sum_k exp(\frac{\hat{C}_{i,k}}{\tau})}, \tag{2.75}$$

where k represents the kth topic, and $\hat{\mathbf{C}}$ represents the initial relationship matrix with a temperature. Assuming that each topic has one and only one parent node, we expect the adjacency matrix $\mathbf{C}$ to be a sparse 0–1 matrix, with at most one value of 1 in each row, and the rest values should be close to 0.

In addition, it is found that if the value matrix in the self-attentive mechanism is introduced, the generated topics will converge. Therefore, unlike the relevance formula in the original self-attentive mechanism, we do not introduce the value matrix when constructing the adjacency matrix. To ensure that each element $C_{i,j}$ is discrete, a softmax function with temperature τ is introduced: For Eq. (2.75), when $\tau \to 0$, it turns C_k into a one-hot vector infinitely close to discrete [28], thus ensuring that each topic has one and only one parent node.

By continuously decreasing the temperature coefficient τ during the training process, the adjacency matrix gradually becomes discrete from the original random, thus clarifying the affiliation between topics. In particular, a topic t_i will be considered as the root topic of a topic tree if the value of $C_{i,i}$ is 1, which means the parent node of a topic is itself.

Topic Decomposition From the perspective of the probabilistic model, it can be considered that the topic-word distribution of the parent topic can be approximated as the sum of the subject word distributions of all its child topics. Based on this principle, the parent and child topics can be related by an adjacency matrix. The global topic distribution and weighted adjacency matrix can be combined to calculate the document-topic distribution $\boldsymbol{\pi}'$ of all parent topics by $\boldsymbol{\pi}' = \boldsymbol{\pi} \times \mathbf{C}$.

After obtaining the representation of the parent topic distribution, we integrate $\boldsymbol{\pi}'$ and $\boldsymbol{\pi}$ to generate the document-topic distribution $\boldsymbol{\theta}$ that merges the parent and child topics as a new topic distribution with mixture level, by $\boldsymbol{\theta} = (1-\gamma)\boldsymbol{\pi}' + \gamma\boldsymbol{\pi}$, where γ is the attenuation coefficient. It should be noted that since the weighted adjacency matrix $\mathbf{C}$ does not capture the correlation between the parent and child topics at the beginning, the original topic distribution is first trained by making $\gamma = 1$. With the continuous learning of the topic self-attention module, the quality of the weighted adjacency matrix $\mathbf{C}$ steadily improves. After $\mathbf{C}$ captures a more reasonable topic parent-child relationship, the value of γ is continuously reduced so that $\boldsymbol{\pi}$ can be gradually replaced by $\boldsymbol{\pi}'$ to learn a well-structured topic forest.

Topic Forest Construction We believe that the hierarchy of topics should be a forest structure, and each tree should be free and flexible. As mentioned above, to construct a reasonable topic forest, we introduce the weighted adjacency matrix $\mathbf{C}$ to implicitly construct the topic forest, where $C_{i,j}$ represents the parent-child relationship between topics. According to the assumption of sparse adjacency matrix, each topic will only be connected to one parent topic. When a topic is connected to itself, it becomes the root of a topic tree. One or more nodes may be connected to it in the adjacency matrix, and thereby modeling the global topic relationship into one or more topic trees can finally form a topic forest.

Forest Constraints We find the structure of the weighted adjacency matrix $\mathbf{C}$ generated by random training probably not a reasonable hierarchical structure. For example, the graph structure constructed by the adjacency matrix may have loops, making it impossible to build a topic forest. It is worth noting that the tree structure itself belongs to a unique DAG, and the topic forest composed of multiple trees also belongs to the DAG.

Therefore, we propose introducing the topic of the parent-child relationship by introducing external acyclic constraints to build it a sparse DAG. Since automatically mining and learning a DAG structure is an NP-hard problem, we introduce the acyclic constraints in [95] to overcome this issue. It is proved that the weighted adjacency matrix $\mathbf{C}$ is a DAG if and only if

$$h(\mathbf{C}) = \mathrm{tr}\left(e^{\mathbf{C} \circ \mathbf{C}}\right) - d = 0, \tag{2.76}$$

where $\circ$ is the Hadamard product and $e^{\mathbf{C}}$ is the matrix exponent of $\mathbf{C}$. Based on this theory, the acyclic property of the adjacency matrix is ensured by introducing the DAG constraint $h(\mathbf{C}) = 0$.

Hierarchal Loss Furthermore, given that the semantic scope of a parent topic should be broader than that of its child topic, it is reasonable to assume that a parent topic can cover all of its child topics. Accordingly, it can be assumed that the document-topic distribution of any parenting topic is similar to the sum of the document-topic distributions of all subtopics corresponding to the topic. That is, a small number of topics in the upper layer can approximately replace most topics in the lower layer and can be used to reconstruct the original document $\boldsymbol{d}$ independently. We refactor the document $\boldsymbol{\pi} \times \boldsymbol{\Phi}$ under the constraint of $h(\mathbf{C}) = 0$ and use the parent document-topic distribution to reconstruct the document $\boldsymbol{\pi}' \times \boldsymbol{\Phi}$ to learn a topic forest, which can be formulated as follows:

$$
\begin{aligned}
&\min_{\mathbf{C}\in\mathbb{R}^{S\times S}} \mathcal{L}_C = \frac{1}{2}\|(\boldsymbol{\pi} - \boldsymbol{\pi}') \times \boldsymbol{\Phi}\|_F^2 \\
&\text{s.t.} h(\boldsymbol{C}) = 0.
\end{aligned}
\tag{2.77}
$$

Augmented Lagrangian Considering that the optimization problem with constraints cannot be solved directly, it is necessary to convert it into a non-convex regular term by constructing an enhanced Lagrangian function and then use the gradient descent method to solve it:

$$
\min_{\mathbf{C}\in\mathbb{R}^{S\times S}} \mathcal{L}_C = \frac{1}{2}\|(\boldsymbol{\pi} - \boldsymbol{\pi}') \times \boldsymbol{\Phi}\|_F^2 + \frac{\rho}{2}|h(\mathbf{C})|^2 + \epsilon h(\mathbf{C}),
\tag{2.78}
$$

where ρ is the penalty coefficient and ϵ is the Lagrange multiplier.

BFS Construction After training, we obtain a sparse adjacency matrix $\boldsymbol{C}$. On this basis, the breadth-first search algorithm is applied to construct the topic forest:

1. It extracts the topic R corresponding to all points whose diagonal value is 1 from $\boldsymbol{C}$ and defines it as the root topic node of all topic trees.
2. For the root node R_i of any topic tree, find all topics P_i with R_i as the parent node through $\boldsymbol{C}$.
3. For the nodes in P_i, also use $\boldsymbol{C}$ to find the topic P_{i+1} corresponding to the next layer.
4. Repeat process 2 and 3 until all topic trees are constructed.

Objective Function Based on the Wasserstein distance of adversarial training [2], we define the objective function as follows:

$$
\begin{aligned}
\min_{G,E}\max_{D} V(D, E, G) = \mathbb{E}_{\boldsymbol{d}\sim p_d}[\underbrace{\mathbb{E}_{\tilde{\boldsymbol{\pi}}\sim p_E(\cdot|\boldsymbol{d})}[D(\boldsymbol{d}, \tilde{\boldsymbol{\pi}})]}_{D(\boldsymbol{d},E(\boldsymbol{d}))}] \\
- \mathbb{E}_{\boldsymbol{\pi}\sim Dir(\alpha)}[\underbrace{\mathbb{E}_{\tilde{\boldsymbol{d}}\sim p_G(\cdot|\boldsymbol{\pi})}[D(\tilde{\boldsymbol{d}}, \boldsymbol{\pi})]}_{D(G(\boldsymbol{\pi}),\boldsymbol{\pi})}].
\end{aligned}
\tag{2.79}
$$

We also define the loss function for the three modules according to the objective function, respectively. In addition, the acyclic constraint $\mathcal{L}_C$ needs to be introduced to construct the topic forest.

For the discriminator, we employ the Wasserstein distance to measure the distance between the distribution of real and fake samples, and the corresponding loss function $\mathcal{L}_D$ is defined as follows:

$$
\mathcal{L}_D = \mathbb{E}_{\boldsymbol{\pi}\sim Dir(\alpha)}[D(G(\boldsymbol{\pi}), \boldsymbol{\pi})] - \mathbb{E}_{\boldsymbol{d}\sim p_d}[D(\boldsymbol{d}, E(\boldsymbol{d}))].
\tag{2.80}
$$

The encoder needs to learn the mapping from document distribution to topic distribution, and its optimization direction is consistent with that of the discriminator.

Therefore, the corresponding loss function is defined as follows:

$$\mathcal{L}_E = \mathbb{E}_{\boldsymbol{d} \sim p_d}[D(\boldsymbol{d}, E(\boldsymbol{d}))]. \tag{2.81}$$

As a module against the encoder, the generator needs to learn the mapping from topic distribution to the document distribution and conduct adversarial learning with the discriminator. The corresponding loss function is defined as follows:

$$\mathcal{L}_G = -\mathbb{E}_{\boldsymbol{\pi} \sim Dir(\alpha)}[D(G(\boldsymbol{\pi}), \boldsymbol{\pi})] + \mathcal{L}_C. \tag{2.82}$$

2.2.4.3 Experimental Results

To verify the effect of our FANTM on the task of hierarchical topic modeling, we evaluate FANTM on several widely used datasets with a manually curated hierarchy of categories. We compare FANTM with representative baselines, using multiple metrics from topic quality, topic hierarchical affinity, and the reasonability of topic structure.

Datasets To effectively evaluate the hierarchical topic model proposed in this study, we employ three datasets: 20News, Reuters, and Wikitext-103, which have been observed to have hierarchical structures in [15] for experiments. The statistics of these datasets are shown in Table 2.21.

Baselines The baseline models include four tree-like hierarchical topic models, three DAG-structured topic models, and nine flat topic models for a comprehensive comparison. The four tree-like hierarchical topic models are hLDA [4], rCRP [38], TSNTM [35], and nTSNTM [15]. The three hierarchical topic models based on the DAG structure are hPAM [55], HSOC [49], and CluHTM [79]. The selected nine flat topic models are GSM [54], GSB [54], NB-NTM [85], GNB-NTM [85], HDP [75], iTM-VAE [62], HiTM-VAE [62], ATM [83], and BAT [84]. We use the open-source codes for all the baselines mentioned above, and the corresponding URLs are marked in the footnotes. Models without open-source code are implemented according to their original papers.

Since the confrontation process of GAN is a binary classification problem for the discriminator, a Sigmod function is added to the last layer of the discriminator, and use the Jensen-Shannon (JS) divergence as a metric. However, many studies [51, 63] have pointed out that the Sigmod function will make the output polarized

Table 2.21 The statistics of datasets

Dataset	# Docs (Train)	#Docs (Test)	Vocabulary size
20News	11,314	7531	1995
Reuters	7769	3019	2000
Wikitext-103	28,472	120	20,000

between [0, 1], making the discrimination cannot distinguish the difference between samples well, because the output given by the discriminator is 1 for all real samples, and the input given by the discriminator is 0 for all fake samples. This results in no intersection between true and false sample distributions, from which the generator cannot learn and update parameters. Therefore, we refer to WGAN [2], which removes the Sigmod function of the last layer of the discriminator and introduces the Wasserstein distance to measure the distance between the two distributions.

Experiment Settings Considering that the inconsistency in the number of topics between parametric and nonparametric topic models may lead to unfair comparison of results, in order to better compare parametric and nonparametric topic models, the number of topics for all parametric models was set to 50 and 200, respectively, for comparison with reference to previous studies [15].

For the nonparametric models based on the folding stick process (i.e., iTM-VAE and nTSNTM), we refer to the original paper to set the maximum number of topics to 200, and the central parameter $\hat{\boldsymbol{\beta}}$ of the GEM distribution is verified according to validation set with the parameter search range [5, 10, 15, 20, 25, 30]. In addition, to keep the nonparametric settings consistent with previous research, this study refers to the practice of nTSNTM [15]. It selects topics with a cumulative average total probability of more than 95% in the GEM distribution as active topics to determine the number of topics s and the corresponding inequality by $\sum_{i=1}^{s} \pi_i \geq 0.95$.

For FANTM, the hidden layer structure in the encoder and discriminator is uniformly set to 150, the learning rate of the discriminator lr_1 is set to 1×10^{-5}, the learning rate of the generator and encoder lr_2 is set to 2×10^{-5}, and the dimension of the topic embedding T_E is set to 300.

For the topic model based on matrix factorization, the default settings of the code implementation in the original paper are followed, and the maximum depth of the topic forest is also set to 3.

Topic Interpretability Topic coherence refers to the degree of correlation between subject words within the same topic and is often used to measure the quality of the topics mined. In addition to manual scoring evaluation, there are various commonly used metrics for topic coherence evaluation [69] according to the selection of word co-occurrence windows. We use Normalized Pointwise Mutual Information (NPMI) scores here [43] to assess topic coherence and interpretability, as this metric has been shown to be close to human judgment [43]. The metric is calculated as follows:

$$\mathrm{NPMI}\left(w_i, w_j\right) = \frac{\log \frac{P(w_i, w_j)}{P(w_i)P(w_j)}}{-\log\left(P\left(w_i, w_j\right)\right)}. \tag{2.83}$$

For the captured topic-word distributions, the top-5, -10, and -15 words with the highest probability of occurrence in each topic are extracted to calculate the average NPMI scores of all the models on the three datasets. For any two words w_i and w_j, the corresponding NPMI score is shown in Table 2.22.

Table 2.22 Average NPMI scores using top-5, -10, and -15 words in each topic

Dataset	20NEWS		Reuters		Wikitext-103	
Number of topics	50	200	50	200	50	200
GSM	0.211	0.165	0.198	0.155	0.214	0.217
GSB	0.231	0.191	0.152	0.136	0.229	0.131
NB-NTM	0.188	0.223	0.248	0.245	0.127	0.125
GNB-NTM	0.240	0.228	0.237	0.255	0.127	0.093
ATM	0.189	0.174	0.238	0.229	0.251	0.248
BAT	0.238	0.228	0.266	0.236	0.026	0.024
HDP	0.192		0.266		0.157	
iTM-VAE	0.195		0.201		0.184	
HiTM-VAE	0.237		0.269		0.233	
rCRP	0.186		0.206		0.201	
hLDA	0.221		0.185		0.186	
TSNTM	0.212		0.206		0.213	
nTSNTM	0.219		0.234		0.237	
hPAM	0.213		0.229		0.223	
HSOC	0.223		0.210		–	
CluHTM	0.219		0.161		–	
FANTM	**0.310**	**0.313**	**0.330**	**0.316**	**0.336**	**0.361**

We can conclude from Table 2.22 that our FANTM has achieved the highest NPMI score in all three datasets, which is significantly improved compared with other models. The real-world dataset does not have a preset corresponding topic hierarchy, so the forest structure in this study can be more adapted to the dataset, thus generating high-quality topics. The results also prove that the strong generative ability of GAN helps the model to extract high-quality topics. In addition, the BAT model trained on the Wikitext-103 dataset suffers from mode collapse, which results in a very low topic coherence. It is probably due to the large size of the vocabulary of this dataset and the training method used by BAT. The weight clipping method [2] causes the gradient to disappear, making it difficult to train the model.

In conclusion, FANTM can build a flexible topic forest which is more in line with the real topic structure of the corpus while the existing topic models can only construct a fixed topic structure, such as a three-layer topic tree.

Parent-Child Topic Interpretability A reasonable topic hierarchy should ensure a strong semantic connection between parent and child topics and maintain a certain degree of independence between non-parent and child topic nodes. The number of topics should also be moderate. If the topics are broad, too many topics can cause the data to be over-clustered into redundant, highly similar topics. In order to evaluate the hierarchical topic model comprehensively and scientifically, three evaluation metrics are used: cross-layer pointwise mutual information [15], hierarchical affinity Harmony [38], and topic uniqueness [61]. The two-level topic models proposed in this study and other baseline models are evaluated from four perspectives: topic-level rationality, parent-child topic coherence, topic redundancy, and topic-level affinity.

Previous topic coherence metrics, such as NPMI, can only measure the quality of a single topic. In order to measure the rationality of the parent-child dependency between topics, a new metric, Cross-Level Normalized Pointwise Mutual Information (CLNPMI) [15], is adopted here to measure the coherence between parent and child topics. The CLNPMI is calculated as follows:

$$\mathrm{CLNPMI}\left(W_p, W_c\right) = \sum_{w_i \in W_p'} \sum_{w_j \in W_c'} \frac{\mathrm{NPMI}\left(w_i, w_j\right)}{\left|W_p'\right| \left|W_c'\right|}, \tag{2.84}$$

where $W_p' = W_p - W_c$ and $W_c' = W_c - W_p$, and W_p and W_c represent the top-N words of a parent topic and one of its children, respectively.

The result is shown in Table 2.23. It is noted here that neither HSOC nor CluHTM converge on Wikitext-103 within 24 hours, so we do not include results for either model. Our FANTM achieves the highest CLNPMI scores on the three datasets, which proves that the self-attention mechanism in FANTM can effectively mine the relationship of topics from the corpus.

Topic Redundancy A reasonable hierarchy should not have too much redundancy while maintaining the similarity between topics. Otherwise, it will lead to too similar topics and a lack of uniqueness. In order to evaluate the redundancy of the topic generated by the topic model, the topic uniqueness (TU) metric [61] is introduced, calculated by $\mathrm{TU} = \mathrm{Count}(\mathrm{Set}(W_{topN}))/(N \times k)$, where $\mathrm{Count}(\mathrm{Set}(W_{topN}))$ represents the number of distinct words in the top-N words of all topics, and k is the number of topics. A higher TU means that more diverse topics are generated.

For all hierarchical topic models, the evaluation method ignores the hierarchical relationship between topics and calculates the average TU value of all topics. In addition, considering that FANTM does not introduce a nonparametric process, for the sake of fairness, when calculating the TU scores, we take a similar average value based on the number determined by the nonparametric topic model as a comparison. It can be seen from Table 2.23 that the TU score of nTSNTM on the Reuters and

Table 2.23 Average CLNPMI and TU scores for hierarchical topic models using top-5, -10, and -15 words in each topic

	20News		Reuters		Wikitext-103	
Model	CLNPMI	TU	CLNPMI	TU	CLNPMI	TU
hLDA	0.065	0.051	0.050	0.447	0.063	0.597
rCRP	0.098	0.285	0.072	0.227	0.088	0.355
TSNTM	0.086	0.430	0.027	0.370	0.065	0.615
nSTNTM	0.109	0.745	0.102	**0.708**	0.113	**0.730**
hPAM	0.046	0.606	0.011	0.470	0.047	0.713
HSOC	0.128	0.231	0.047	0.211	–	–
CluHTM	0.123	0.116	0.016	0.117	–	–
FANTM	**0.152**	**0.776**	**0.111**	0.677	**0.152**	0.582

Wikitext-103 datasets performs the best, which may be due to the nonparametric process introduced by it that can automatically determine the appropriate number of topics to avoid topic redundancy.

Table 2.23 shows the TU results of all the models. Higher TU values indicate better performances, and the highest scores are marked in bold. It can be observed from Table 2.23 that FANTM can always generate more diverse (i.e., less redundant) topics than other models. Due to the limitation of probability map-based models, the topics learned by hLDA are very similar, resulting in a low TU value. In addition, it is noted that the TU scores of two nonnegative matrix factorization-based DAG-structured topic models, HSOC and CluHTM, are very low. Because they both need to preset a certain amount of topic trees to avoid missing key topic-level relationships. Hundreds of topics lead to excessive redundancy.

Topic Hierarchical Affinity A reasonable assumption about topic hierarchies is that topics forming parent-child relationships exhibit greater similarity in topic-word distribution than non-parent-child topics [38]. Referring to [38], we introduce topic affinity to measure the similarity between parent and child topics. Let $\mathbf{\Phi}(k)$ be the topics at layer k, $\lambda(k)$ be the set of subtopics of $\mathbf{\Phi}(k)$, $\bar{\lambda}(k)$ be the set of non-child topics of $\mathbf{\Phi}(k)$, i.e., topics with no parent-child relationship. To measure topic affinity, we compare the average cosine similarity $S_{\lambda}(k)$ between all topics in $\mathbf{\Phi}(k)$ and $\lambda(k)$, with the average cosine similarity $S_{\bar{\lambda}}(k)$ between all topics in $\mathbf{\Phi}(k)$ and $\bar{\lambda}(k)$. The difference between $S_{\lambda}(k)$ and $S_{\bar{\lambda}}(k)$ represents the difference between parent-child similarity and non-parent-child similarity, and the relevant calculation is as follows:

$$\begin{aligned} S_{\lambda}(k) &= \frac{1}{|\lambda(k)|} \sum_{i \in \lambda(k)} \frac{\phi_k \cdot \phi_i}{|\phi_k|\,|\phi_i|}, \\ S_{\bar{\lambda}}(k) &= \frac{1}{\left|\bar{\lambda}(k)\right|} \sum_{i \in \bar{\lambda}(k)} \frac{\phi_k \cdot \phi_i}{|\phi_k|\,|\phi_i|}. \end{aligned} \tag{2.85}$$

Figure 2.35 shows the corresponding hierarchical affinity score values in all hierarchical topic models. The greater the score difference between parent-child topics and non-parent-child topics is, the better the performance is. Since all topics between the two levels in hPAM are fully connected, parent-child and non-parent-child relationships between topics cannot be distinguished, so hPAM is excluded from this part of experiments. Note that since both CluHTM and HSOC are based on nonnegative matrix factorization, they rely on a predefined number of topic trees. They must set a large number of topic trees (e.g., 100) to avoid missing potential topic hierarchies. However, presetting too many topic trees will lead to an exponential increase in the size of non-subtopic sets, so the average $S_{\bar{\lambda}}(k)$ corresponding to these two models will be lower, resulting in more competitive topic affinity.

There exists a trade-off that too many topic trees and topics can also bring high topic redundancy for CluHTM and HSOC, as shown by the TU value in Table 2.23.

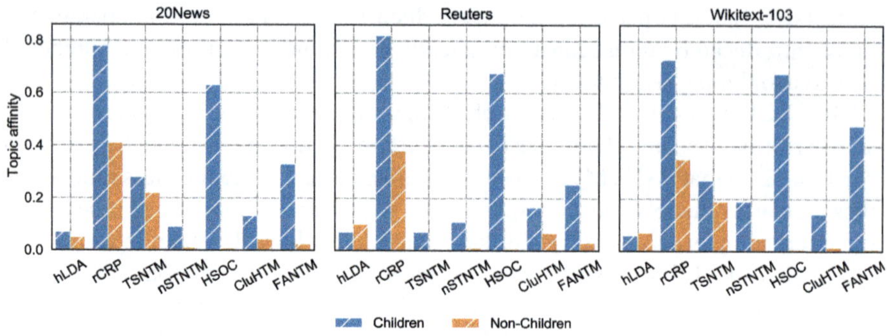

Fig. 2.35 Topic affinity scores of all models showing distinct differences between subtopics (blue and orange columns)

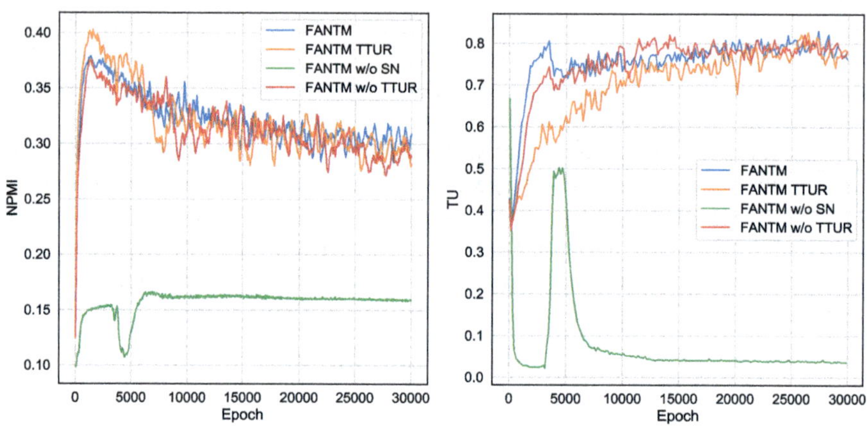

Fig. 2.36 The influence of SN and TTUR on GAN model in 20News training process

Ablation Study We refer to previous studies and introduce two methods: spectral norm normalization and dual time-scale update rules to ensure the stability of adversarial training, so that the discriminator can satisfy the Lipschitz continuity. Under this condition, we maintain a faster training speed than the generator to guide the generator for stable learning. In this part, we conduct ablation experiments on these two modules on the 20News dataset and evaluate the impact of these two methods on the stability of model training through the index changes during the training process.

Figure 2.36 shows the results of the ablation experiments in this part. FANTM w/o SN represents the GAN-based model in this study which removes the spectral norm normalization module and uses the Weight Clipping method to limit the gradient backhaul, while FANTM w/o TTUR represents the GAN-based model in this study which removes the dual time-scale update rules and uses previous methods of training the discriminator for multiple rounds. The experimental results show that after deleting the SN module, FANTM cannot be trained. However, it can

Table 2.24 The effect of different numbers of topics on the performance of FANTM over the 20News dataset

Topic number	NPMI	TU	CLNPMI
50	0.310	**0.812**	0.134
100	0.311	0.652	**0.171**
200	**0.313**	0.514	0.152

still show a certain learning ability in the first few iterations. But then there were big fluctuations, in which the TU value also dropped sharply, indicating a mode collapse [45]. The removal of TTUR has no significant effect on the overall training, and the curve is also a large fluctuation.

Hyperparameter Evaluation For the hierarchical topic model based on GAN, we explore the influence of the number of preset topics on FANTM over the 20News dataset. The corresponding results are shown in Table 2.24. It can be found from Table 2.24 that as the number of topics increases, the NPMI value of the model does not fluctuate significantly, while the TU value decreases significantly with the increase of the number of topics. CLNPMI presents a trend of rising first and then falling, which indicates that too many or too few topics will lead to a decrease in the consistency between parent and child topics.

In addition, we also studied the effect of the temperature τ on the generation of adjacency matrix. Please refer to the Appendix for details.

Temperature Introducing a softmax function with a temperature value parameter τ can help build a sparse adjacency matrix. In this part the evolution of the adjacency matrix with decreasing temperature will be shown. Since the model in this study will exponentially update the temperature value every multiple training rounds during the training process, sampling is performed after each temperature change, and the corresponding adjacency matrix is saved. Then we visualize it through a heatmap.

Figure 2.37 shows the adjacency matrix heatmap of FANTM on 20News when the temperature value continues to decrease. The order of change is from left to right and from top to bottom. It can be found from Fig. 2.37 that as the temperature value changes from high to low, the value of the adjacency matrix gradually becomes discretized from a uniform distribution until it finally presents a black and white 0–1 distribution. This change shows that there is no clear affiliation between themes at the beginning. As the temperature decreases, the affiliation between themes is gradually clarified, and finally, the parent-child relationship between all themes can be described.

Visualization We here visualize the generated adjacency matrix to show the effect of FANTM. We use Gephi[7] to achieve this automatically. The weighted adjacency matrices are shown in Fig. 2.38. The values of the white dots represent the two topics. There is a parent-child relationship, for example, (0, 42) means that topic 42 is the parent node of topic 0, and the dots on the diagonal represent the root topic.

[7] http://gephi.org/

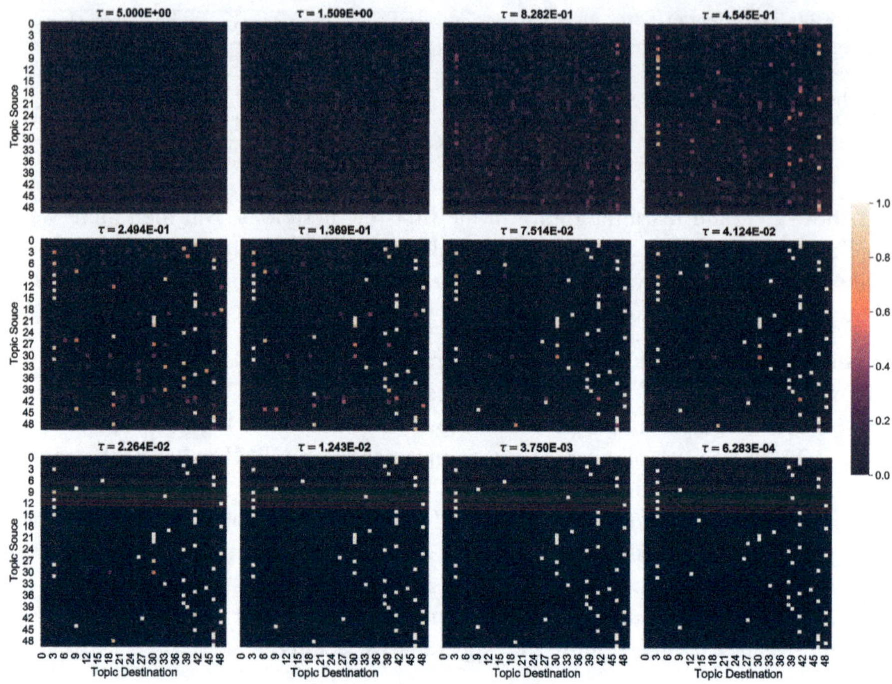

Fig. 2.37 The effect of temperature value τ on FANTM generating sparse adjacency matrix on 20News

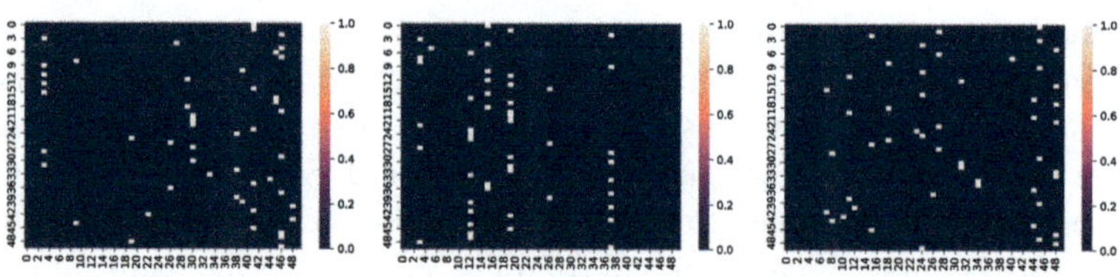

Fig. 2.38 Visualization of the weighted adjacency matrices generated from 20 ns (left), Reuters (medium), and Wikitext-103 (right)

The rest of the points are discretely distributed and satisfy the acyclic constraint, except for the points on the diagonal.

Topic Forest To further describe the topic forest structure, we show the global topic forests generated by the model on 20News, Reuters, and Wikitext-103 in Figs. 2.39, 2.40, and 2.41, respectively.

In the generated topic forest, each node represents a topic. The first five topic words form node labels. The arrow direction is from the parent topic to the child topic. The color of each node is determined by the degree of the node.

From Figs. 2.39, 2.40, and 2.41, it can be found that the underlying topic forest structure of different datasets is very different. For example, the topic forest

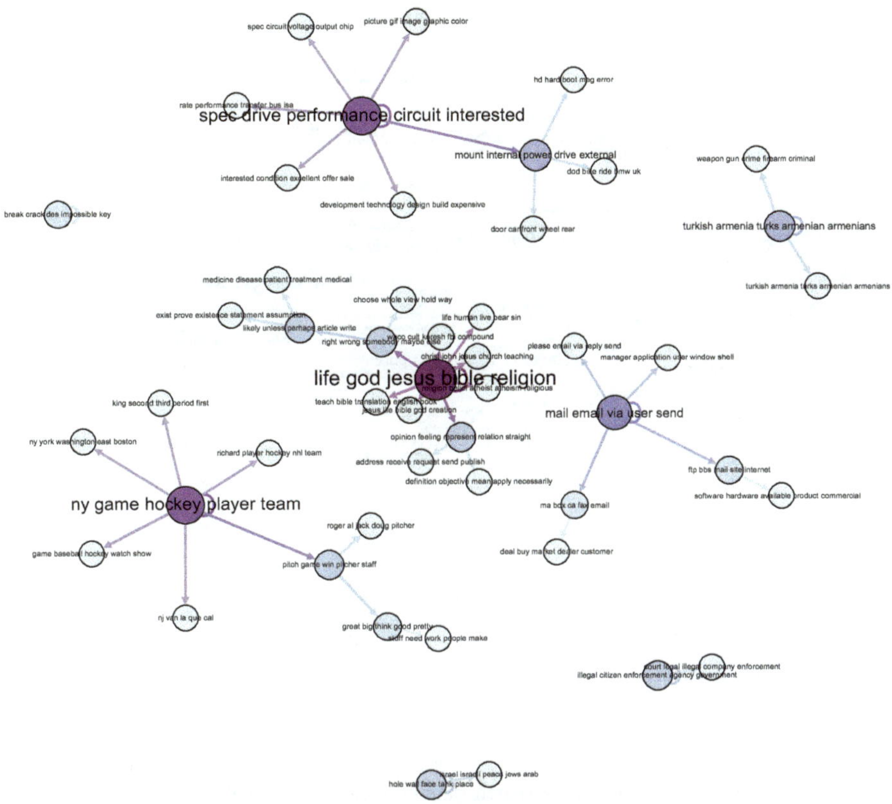

Fig. 2.39 Topic forest generated from 20News dataset. It is obvious that the topics related to *religion* are clustered in the middle tree, and the topics related to *computer* and *game* are clustered in other trees

generated in the Reuters dataset contains the main tree whose root topic is about news interviews. The topic forests generated by the model in this study on 20News and Wikitext-103 are composed of multiple trees. The root topics mainly involve news and daily life, respectively. The topic trees in Wikitext-103 are mostly shallow and scattered, indicating that this dataset contains many independent topics, which aligns with an encyclopedia's data distribution characteristics. All results show that the topic hierarchy obtained by the model is reasonable, e.g., the root topic is generic and the subtopics are highly related to their corresponding parent topics.

Generally, FANTM can build a flexible topic forest which is more in line with the real topic structure of the corpus, while the existing topic models can only construct a fixed topic structure, such as a three-layer topic tree. Empirical results show that FANTM outperforms the baseline by 29.17% in NPMI and reaches state-of-the-art performance in hierarchical metrics such as CLNPMI.

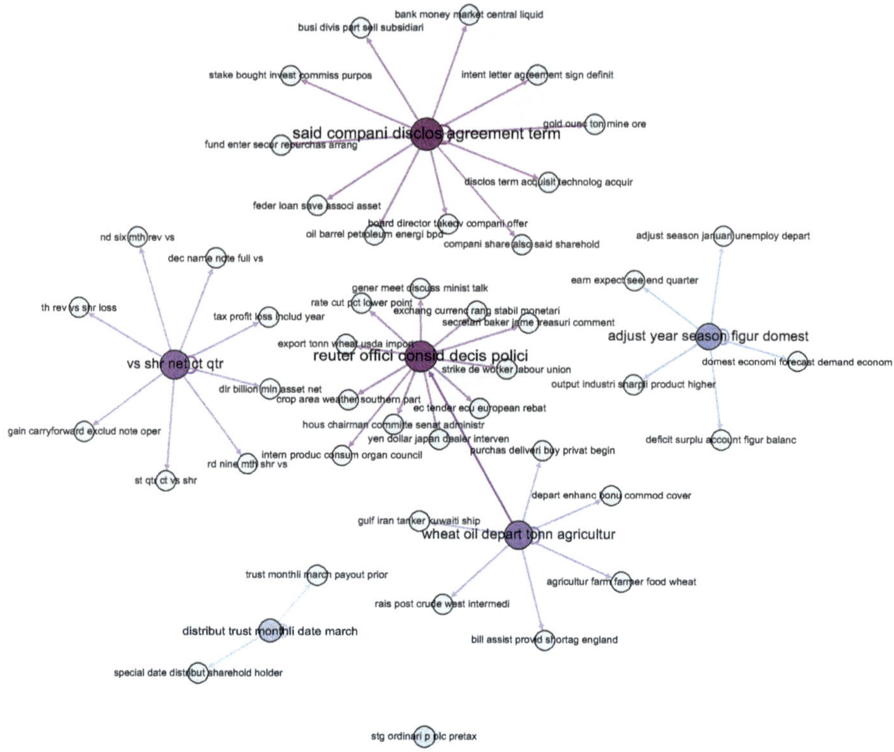

Fig. 2.40 Topic forest generated from Reuters dataset. It can be seen that there are five topic trees in the generated topic forest. There are many topics related to interviews in various fields under the topic of interviews on behalf of Reuters

2.2.5 *Summary of Hierarchical Topic Models*

In summary, the aforementioned works showcase significant advancements in hierarchical topic models for handling hierarchical structures within textual data. From the tree structure of nTSNTM to the forest structure of nFNTM, further to the flexible forest in FANTM, and to the deep generative model of NSEM-GMHTM, these models contribute to capturing complex relationships between topics and enhancing model interpretability and adaptability. The development of these models not only drives theoretical innovation in hierarchical topic models but also provides new tools and methods for text analysis and data exploration in practical applications. With the further development of deep learning and NVI techniques, we anticipate that hierarchical topic models will be able to handle larger datasets and play a role in a wider range of application domains in the future.

Fig. 2.41 Topic forest generated from the Wikitext-103 dataset, illustrating the broad coverage and shallow hierarchical structure of Wikipedia-derived topic trees

References

1. Alvarez-Melis D, Jaakkola TS (2017) Tree-structured decoding with doubly-recurrent neural networks. In: Proceedings of the 5th international conference on learning representations
2. Arjovsky M, Chintala S, Bottou L (2017) Wasserstein generative adversarial networks. In: Proceedings of the 34th international conference on machine learning, pp 214–223
3. Blei DM, McAuliffe JD (2007) Supervised topic models. In: Advances in neural information processing systems, pp 121–128
4. Blei DM, Griffiths TL, Jordan MI, Tenenbaum JB (2003) Hierarchical topic models and the nested Chinese restaurant process. In: Advances in neural information processing systems
5. Blei DM, Ng AY, Jordan MI (2003) Latent Dirichlet allocation. J Mach Learn Res 3:993–1022
6. Blei DM, Griffiths TL, Jordan MI (2010) The nested Chinese restaurant process and Bayesian nonparametric inference of topic hierarchies. J ACM 57(2):7:1–7:30
7. Blondel VD, Guillaume JL, Lambiotte R, Lefebvre E (2008) Fast unfolding of communities in large networks. J Stat Mech Theory Experiment 2008(10):P10008
8. Bollen KA (1989) Structural equations with latent variables. John Wiley & Sons, London
9. Bouma G (2009) Normalized (pointwise) mutual information in collocation extraction. In: Proceedings of German society for computational linguistics, pp 31–40
10. Burkhardt S, Kramer S (2019) Decoupling sparsity and smoothness in the Dirichlet variational autoencoder topic model. J Mach Learn Res 20(131):1–27

11. Cao Z, Li S, Liu Y, Li W, Ji H (2015) A novel neural topic model and its supervised extension. In: Proceedings of the 29th AAAI conference on artificial intelligence, pp 2210–2216
12. Card D, Tan C, Smith NA (2018) Neural models for documents with metadata. In: Proceedings of the 56th annual meeting of the association for computational linguistics, pp 2031–2040
13. Chen H, Mao P, Lu Y, Rao Y (2023) Nonlinear structural equation model guided Gaussian mixture hierarchical topic modeling. In: Proceedings of the 61st annual meeting of the association for computational linguistics, pp 10377–10390
14. Chen Z, Ding C, Rao Y, Xie H, Tao X, Cheng G, Wang FL (2021) Hierarchical neural topic modeling with manifold regularization. World Wide Web 24:2139–2160
15. Chen Z, Ding C, Zhang Z, Rao Y, Xie H (2021) Tree-structured topic modeling with nonparametric neural variational inference. In: Proceedings of the 59th annual meeting of the association for computational linguistics and the 11th international joint conference on natural language processing, pp 2342–2353
16. Cheng D, Liu Y (2014) Parallel Gibbs sampling for hierarchical Dirichlet processes via gamma processes equivalence. In: Proceedings of the 20th ACM SIGKDD international conference on knowledge discovery and data mining, pp 562–571
17. Clauset A, Moore C, Newman ME (2008) Hierarchical structure and the prediction of missing links in networks. Nature 453(7191):98–101
18. Dieng AB, Ruiz FJR, Blei DM (2020) Topic modeling in embedding spaces. Trans Assoc Comput Linguist 8:439–453
19. Dilokthanakul N, Mediano PA, Garnelo M, Lee MC, Salimbeni H, Arulkumaran K, Shanahan M (2016) Deep unsupervised clustering with Gaussian mixture variational autoencoders. CoRR abs/1611.02648
20. Donahue J, Krähenbühl P, Darrell T (2017) Adversarial feature learning. In: Proceedings of the 5th international conference on learning representations
21. Duan Z, Wang D, Chen B, Wang C, Chen W, Li Y, Ren J, Zhou M (2021) Sawtooth factorial topic embeddings guided gamma belief network. In: Proceedings of the 38th international conference on machine learning, pp 2903–2913
22. Feng Y, Feng J, Rao Y (2020) Reward-modulated adversarial topic modeling. In: Proceedings of the 25th international conference on database systems for advanced applications, pp 689–697
23. Figurnov M, Mohamed S, Mnih A (2018) Implicit reparameterization gradients. In: Advances in neural information processing systems, pp 439–450
24. Goldberger AS (1972) Structural equation methods in the social sciences. J Econom Soc 40(6):979–1001
25. Green PJ (1995) Reversible jump Markov chain Monte Carlo computation and Bayesian model determination. Biometrika 82(4):711–732
26. Griffiths TL, Steyvers M, Blei DM, Tenenbaum JB (2004) Integrating topics and syntax. In: Advances in neural information processing systems, pp 537–544
27. Hennig P, Stern D, Herbrich R, Graepel T (2012) Kernel topic models. In: Proceedings of the fifteenth international conference on artificial intelligence and statistics, pp 511–519
28. Hinton GE, Vinyals O, Dean J (2015) Distilling the knowledge in a neural network. CoRR abs/1503.02531
29. Hofmann T (1999) Probabilistic latent semantic analysis. In: Proceedings of the 15th conference on uncertainty in artificial intelligence, pp 289–296
30. Hoyle A, Goel P, Hian-Cheong A, Peskov D, Boyd-Graber J, Resnik P (2021) Is automated topic model evaluation broken? The incoherence of coherence. In: Advances in neural information processing systems, pp 2018–2033
31. Hoyle AM, Goel P, Sarkar R, Resnik P (2022) Are neural topic models broken? In: Findings of the association for computational linguistics: EMNLP, pp 5321–5344
32. Hu X, Wang R, Zhou D, Xiong Y (2020) Neural topic modeling with cycle-consistent adversarial training. In: Proceedings of the 2020 conference on empirical methods in natural language processing

33. Huang M, Rao Y, Liu Y, Xie H, Wang FL (2018) Siamese network-based supervised topic modeling. In: Proceedings of the 2018 conference on empirical methods in natural language processing, pp 4652–4662
34. Ishwaran H, James LF (2001) Gibbs sampling methods for stick-breaking priors. J Am Stat Assoc 96(453):161–173
35. Isonuma M, Mori J, Bollegala D, Sakata I (2020) Tree-structured neural topic model. In: Proceedings of the 58th annual meeting of the association for computational linguistics, pp 800–806
36. Jang E, Gu S, Poole B (2017) Categorical reparameterization with Gumbel-softmax. In: Proceedings of the 5th international conference on learning representations
37. Jones M (2009) Kumaraswamy's distribution: a beta-type distribution with some tractability advantages. Stat Methodol (1):70–81
38. Kim JH, Kim D, Kim S, Oh AH (2012) Modeling topic hierarchies with the recursive Chinese restaurant process. In: Proceedings of the 21st ACM international conference on information and knowledge management, pp 783–792
39. Kim S, Zhang J, Chen Z, Oh AH, Liu S (2013) A hierarchical aspect-sentiment model for online reviews. In: Proceedings of the AAAI conference on artificial intelligence, pp 526–533
40. Kingma DP, Ba J (2015) Adam: A method for stochastic optimization. In: Proceedings of the 3rd international conference on learning representations
41. Kingma DP, Welling M (2014) Auto-encoding variational Bayes. In: Proceeding of the 2nd international conference on learning representations
42. Kumaraswamy P (1980) A generalized probability density function for double-bounded random processes. J Hydrol 46(1–2):79–88
43. Lau JH, Newman D, Baldwin T (2014) Machine reading tea leaves: Automatically evaluating topic coherence and topic model quality. In: Proceedings of the 14th conference of the European chapter of the association for computational linguistics, pp 530–539
44. Lee DD, Seung HS (1999) Learning the parts of objects by non-negative matrix factorization. Nature 401(6755):788–791
45. Lei N, Guo Y, An D, Qi X, Luo Z, Yau S, Gu X (2019) Mode collapse and regularity of optimal transportation maps. CoRR abs/1902.02934
46. Li W, McCallum A (2006) Pachinko allocation: Dag-structured mixture models of topic correlations. In: Proceedings of the 23th international conference on machine learning, pp 577–584
47. Lin L, Rao Y, Xie H, Lau RYK, Yin J, Wang FL, Li Q (2022) Copula guided parallel Gibbs sampling for nonparametric and coherent topic discovery. IEEE Trans Knowl Data Eng 34(1):219–235
48. Liu L, Huang H, Gao Y, Zhang Y, Wei X (2019) Neural variational correlated topic modeling. In: Proceeding of the world wide web conference, pp 1142–1152
49. Liu R, Wang X, Wang D, Zuo Y, Zhang H, Zheng X (2018) Topic splitting: a hierarchical topic model based on non-negative matrix factorization. J Syst Sci Syst Eng (4):479–496
50. Luo Y, Peng J, Ma J (2020) When causal inference meets deep learning. Nat Mach Intell 2(8):426–427
51. Mao X, Li Q, Xie H, Lau RYK, Wang Z, Smolley SP (2017) Least squares generative adversarial networks. In: Proceedings of the IEEE international conference on computer vision, pp 2794–2802
52. McInnes L, Healy J, Melville J (2018) UMAP: uniform manifold approximation and projection for dimension reduction. CoRR abs/1802.03426
53. Miao Y, Yu L, Blunsom P (2016) Neural variational inference for text processing. In: Proceedings of the 33rd international conference on machine learning, pp 1727–1736
54. Miao Y, Grefenstette E, Blunsom P (2017) Discovering discrete latent topics with neural variational inference. In: Proceedings of the 34th international conference on machine learning, pp 2410–2419
55. Mimno DM, Li W, McCallum A (2007) Mixtures of hierarchical topics with pachinko allocation. In: Proceedings of the 24th international conference on machine learning, pp 633–640

56. Ming Z, Wang K, Chua T (2010) Prototype hierarchy based clustering for the categorization and navigation of web collections. In: Proceedings of the 33rd international ACM SIGIR conference on research and development in information retrieval, pp 2–9
57. Mnih A, Gregor K (2014) Neural variational inference and learning in belief networks. In: Proceedings of the 31st international conference on machine learning, pp 1791–1799
58. Moreo A, Esuli A, Sebastiani F (2016) Distributional random oversampling for imbalanced text classification. In: Proceedings of the 39th international ACM SIGIR conference on research and development in information retrieval, pp 805–808
59. Naesseth C, Ruiz F, Linderman S, Blei D (2017) Reparameterization gradients through acceptance-rejection sampling algorithms. In: Proceedings of the 20th international conference on artificial intelligence and statistics, pp 489–498
60. Nalisnick ET, Smyth P (2017) Stick-breaking variational autoencoders. In: ICLR 2017
61. Nan F, Ding R, Nallapati R, Xiang B (2019) Topic modeling with Wasserstein autoencoders. In: Proceedings of the 57th annual meeting of the association for computational linguistics, pp 6345–6381
62. Ning X, Zheng Y, Jiang Z, Wang Y, Yang H, Huang J, Zhao P (2020) Nonparametric topic modeling with neural inference. Neurocomputing 399:296–306
63. Nowozin S, Cseke B, Tomioka R (2016) f-GAN: training generative neural samplers using variational divergence minimization. In: Advances in neural information processing systems, pp 271–279
64. Paisley JW, Wang C, Blei DM, Jordan MI (2015) Nested hierarchical Dirichlet processes. IEEE Trans Pattern Anal Mach Intell (2):256–270
65. Pennington J, Socher R, Manning CD (2014) Glove: Global vectors for word representation. In: Proceedings of the 2014 conference on empirical methods in natural language processing, pp 1532–1543
66. Pitman J (2006) Combinatorial stochastic processes
67. Ramage D, Hall D, Nallapati R, Manning CD (2009) Labeled LDA: A supervised topic model for credit attribution in multi-labeled corpora. In: Proceedings of the 2009 conference on empirical methods in natural language processing, pp 248–256
68. Rezende DJ, Mohamed S, Wierstra D (2014) Stochastic backpropagation and approximate inference in deep generative models. In: Proceedings of the 31st international conference on machine learning, pp 1278–1286
69. Röder M, Both A, Hinneburg A (2015) Exploring the space of topic coherence measures. In: Proceedings of the 8th ACM international conference on web search and data mining, pp 399–408
70. Sethuraman J (1994) A constructive definition of Dirichlet priors. Statistica Sinica 639–650
71. Shu H, Zhou J, Lian Q, Li H, Zhao D, Zeng J, Ma J (2021) Modeling gene regulatory networks using neural network architectures. Nat Comput Sci 1(7):491–501
72. Speer R, Chin J, Havasi C (2017) ConceptNet 5.5: An open multilingual graph of general knowledge. In: Proceedings of the 31st AAAI conference on artificial intelligence, pp 4444–4451
73. Srivastava A, Sutton C (2017) Autoencoding variational inference for topic models. In: Proceedings of the 5th international conference on learning representations
74. Tan C, Card D, Smith NA (2017) Friendships, rivalries, and trysts: Characterizing relations between ideas in texts. In: Proceedings of the 55th annual meeting of the association for computational linguistics, pp 773–783
75. Teh YW, Jordan MI, Beal MJ, Blei DM (2004) Sharing clusters among related groups: hierarchical Dirichlet processes. In: Advances in neural information processing systems, pp 1385–1392
76. Titsias M (2007) The infinite Gamma-Poisson feature model. In: Advances in neural information processing systems
77. Vamshi KB, Pandey AK, Siva KAP (2018) Topic model based opinion mining and sentiment analysis. In: Proceedings of the 2018 international conference on computer communication and informatics, pp 1–4

78. Vaswani A, Shazeer N, Parmar N, Uszkoreit J, Jones L, Gomez AN, Kaiser L, Polosukhin I (2017) Attention is all you need. In: Advances in neural information processing systems, pp 5998–6008
79. Viegas F, Cunha W, Gomes C, Pereira A, Rocha L, Gonçalves MA (2020) CluHTM—semantic hierarchical topic modeling based on CluWords. In: Proceedings of the 58th annual meeting of the association for computational linguistics, pp 8138–8150
80. Wainwright MJ, Jordan MI, et al. (2008) Graphical models, exponential families, and variational inference. Found Trends® Mach Learn 1(1–2):1–305
81. Wang C, Blei DM, Fei-Fei L (2009) Simultaneous image classification and annotation. In: Proceedings of the IEEE conference on computer vision and pattern recognition, pp 1903–1910
82. Wang D, Guo D, Zhao H, Zheng H, Tanwisuth K, Chen B, Zhou M (2022) Representing mixtures of word embeddings with mixtures of topic embeddings. In: Proceedings of the 10th international conference on learning representations
83. Wang R, Zhou D, He Y (2019) ATM: adversarial-neural topic model. Information Process Manag 56(6):102098
84. Wang R, Hu X, Zhou D, He Y, Xiong Y, Ye C, Xu H (2020) Neural topic modeling with bidirectional adversarial training. In: Proceedings of the 58th annual meeting of the association for computational linguistics, pp 340–350
85. Wu J, Rao Y, Zhang Z, Xie H, Li Q, Wang FL, Chen Z (2020) Neural mixed counting models for dispersed topic discovery. In: Proceedings of the 58th annual meeting of the association for computational linguistics, pp 6159–6169
86. Wu X, Dong X, Nguyen TT, Luu AT (2023) Effective neural topic modeling with embedding clustering regularization. In: Proceedings of the 40th international conference on machine learning, pp 37335–37357
87. Xiong L, Xu K, Tian K, Shao Y, Tang L, Gao G, Zhang M, Jiang T, Zhang QC (2019) Scale method for single-cell ATAC-seq analysis via latent feature extraction. Nat Commun 10(1):4576
88. Xu Y, Wang D, Chen B, Lu R, Duan Z, Zhou M (2022) Hyperminer: Topic taxonomy mining with hyperbolic embedding. In: Advances in neural information processing systems, pp 31557–31570
89. Xue F, Hong R, He X, Wang J, Qian S, Xu C (2020) Knowledge-based topic model for multi-modal social event analysis. IEEE Trans Multimedia 22(8):2098–2110
90. Yang L, Cheung NM, Li J, Fang J (2019) Deep clustering by Gaussian mixture variational autoencoders with graph embedding. In: Proceedings of the IEEE/CVF international conference on computer vision, pp 6440–6449
91. Yao S, Peng B, Papadimitriou CH, Narasimhan K (2021) Self-attention networks can process bounded hierarchical languages. In: Proceedings of the 59th annual meeting of the association for computational linguistics and the 11th international joint conference on natural language processing, pp 3770–3785
92. Yu Y, Chen J, Gao T, Yu M (2019) DAG-GNN: DAG structure learning with graph neural networks. In: Proceedings of the 36th international conference on machine learning, pp 7154–7163
93. Zhang Z, Zhang X, Rao Y (2022) Nonparametric forest-structured neural topic modeling. In: Proceedings of the 29th international conference on computational linguistics, pp 2585–2597
94. Zhao H, Phung D, Huynh V, Le T, Buntine W (2021) Neural topic model via optimal transport. In: Proceedings of the 9th international conference on learning representations
95. Zheng X, Aragam B, Ravikumar P, Xing EP (2018) Dags with NO TEARS: continuous optimization for structure learning. In: Advances in neural information processing systems, pp 9492–9503
96. Zhou M, Carin L (2012) Augment-and-conquer negative binomial processes. In: Advances in neural information processing systems, pp 2555–2563

Chapter 3
Modern Topic Models

Abstract With the development of user-generated content, many real-world corpora involve various types of metadata, such as timestamp and manually defined labels, and yet most of the classical topic modeling methods ignore this information. Besides, many contemporary topic models continue to grapple with the issue of noise contamination, particularly when they are applied to small or social media datasets. Given a corpus where each document contains a timestamp, lifelong topic models are developed to enhance topic quality over such documents sorted by timestamp, but the order of topics extracted by existing methods may shift sequentially. To address this issue, we exploit nonnegative matrix tri-factorization, a classical method for co-clustering, as a framework of lifelong topic modeling with an aligned topic space. We also develop a data distribution scheme followed by a distributed parallel algorithm, which scales our model to large-scale text streams. In this vein, a lifelong hierarchical topic model based on nonparametric clustering is further proposed. Finally, we develop a label aligned neural topic model, an anchor guided hierarchical topic model, and a self-supervised hierarchical topic model, to alleviate the issue of noise contamination by exploiting labels or embeddings.

In this chapter, we will introduce our work entitled "NMTF-LTM: Towards an Alignment of Semantics for Lifelong Topic Modeling" [38], "Lifelong Hierarchical Topic Modeling via Nonparametric Word Embedding Clustering" [76], "Supervised Neural Topic Modeling with Label Alignment" [11], "Unsupervised Hierarchical Topic Modeling via Anchor Word Clustering and Path Guidance" [44], and "Self-supervised Topic Taxonomy Discovery in the Box Embedding Space" [46]. The above methods, which utilize clustering algorithms, the entropy-regularized optimal transport, or embeddings, constitute the main strategies in modern topic modeling.

Y. Rao, Q. Li, *Topic Modeling*, Machine Learning: Foundations, Methodologies, and Applications, https://doi.org/10.1007/978-981-96-8853-1_3

3.1 Lifelong Topic Models

3.1.1 Aligned Lifelong Topic Model

By accumulating semantic knowledge for future learning, lifelong topic models have shown their advantages in discovering latent topics over a text stream continuously, while most of which suffer from the *semantic misalignment* problem, creating obstacles toward tracking semantic evolution and applying the models to downstream tasks. To address this issue, NMTF-LTM is proposed, which consolidates a semantic alignment by constraining the mapped topic space. An efficient MPI-based parallel algorithm, PNMTF-LTM, is also developed and theoretically analyzed, which is verified and scalable under a large-scale text stream with more than a million documents.

3.1.1.1 Introduction

Most of the existing lifelong topic models are observed to be unable to retain a consistent permutation of topics over a text stream. Specifically, the order of topics in the *word-topic* matrix (i.e., basis) may change at each time. For example, for the *word-topic* basis obtained by NMF-LTM on AGNews at time $t-1$, the first topic represents the semantics of "War," while this topic shifts to the fifth position at time t, as shown in the bottom part of Fig. 3.1.

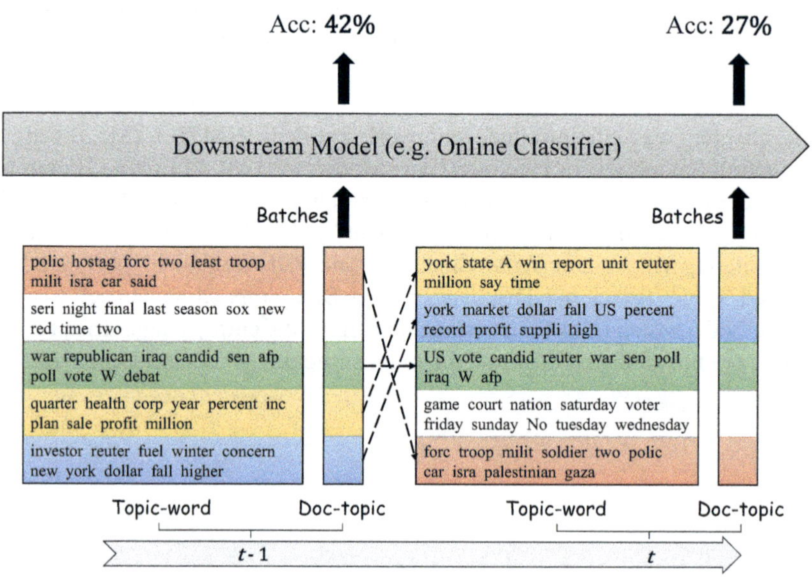

Fig. 3.1 Semantic misalignment example on AGNews

To further demonstrate the above problem, we introduce an example of online classification where the classifier is updated by incoming mini-batches [64]. As shown in the upper part of Fig. 3.1, we regard the *document-topic* representations at time t as the tth mini-batch and use these batches to update the classifier in an online manner. It can be observed that the classification accuracy declines from 42% at time $t-1$ to 27% at time t. The reason for such a deteriorated performance is that the semantic misalignment between the *document-topic* representations of time $t-1$ and time t leads to a too rapid semantic drift, breaking the *i.i.d.* assumption of online classification. Specifically, the classifier was trained with "War" as the first dimension of the input *document-topic* representation at time $t-1$, while this dimension shifts to "Business" at time t. A similar issue also arises on other downstream tasks, such as topic-level IR under the scenario of text stream processing. In this work, we exploit nonnegative matrix tri-factorization (NMTF) [21], i.e., an extension of NMF, as a framework for LTM. By mapping the *word-topic* basis to a new aligned topic space through an extra factor, the new topic representations share aligned semantics for documents. To the best of our knowledge, we are the first to propose and tackle such a *semantic misalignment* problem in the LTM task.

Another major limitation for NMF-based topic models is the high space and time complexity due to large data matrices and intensive matrix multiplications. With the explosion of data volume, there is a growing speed in the generation of data under the stream. The size of data in a given time window, i.e., the size of a data chunk, is becoming larger and larger. Such a large-scale data chunk is infeasible to be processed in time by a single computing machine whose memory and computing resources are limited. Therefore, to meet the real-time requirement of stream computing, it is necessary to design a distributed parallel algorithm for the NMF-based LTM task. Unfortunately, it remains quite challenging in this task to partition and distribute each matrix, as well as coordinate and communicate local computations only with the partitioned matrices.

Toward addressing these issues, we propose an NMTF-LTM method based on NMTF and its enhanced PNMTF-LTM method equipped with a distributed parallel algorithm. The main contributions are summarized as follows:

- We first illustrate the phenomenon of *topic reordering* and the problem of *semantic misalignment* in LTM, as well as emphasize the importance of this problem on downstream tasks.
- We address the *semantic misalignment problem* in LTM by a consolidated regularization employed on the mapped topic space over time.
- We develop a data distribution scheme followed by a distributed parallel algorithm based on the Message Passing Interface (MPI), which scales our model to large-scale text streams.

3.1.1.2 Methodology

Definition 3.1 (Topic Reordering) For two sequences $\mathbf{W}_a = \{w_{a_1}, w_{a_2}, \ldots, w_{a_K}\}$ and $\mathbf{W}_b = \{w_{b_1}, w_{b_2}, \ldots, w_{b_K}\}$ representing topics at time t_a and t_b ($t_a \neq t_b$), respectively, topic reordering happens between t_a and t_b if and only if

$$\begin{aligned} \exists i, j : & \left(s(w_{a_i}) \approx s(w_{b_j})\right) \wedge (i \neq j) \wedge \\ & \left(s(w_{a_i}) \neq s(w_{b_i})\right) \wedge \left(s(w_{a_j}) \neq s(w_{b_j})\right), \end{aligned} \tag{3.1}$$

where $s(w)$ identifies the semantics of topic w.

For two particular topic indices i and j, the first two propositions in Eq. (3.1) describe the case that the ith topic at time t_a and the jth topic at time t_b share the same semantics, while these two topics are not aligned with each other for $i \neq j$. The last two propositions are used to eliminate the special cases of topic redundancy, i.e., $s(w_{a_i}) \approx s(w_{b_j}) \approx s(w_{b_i})$ and $s(w_{a_j}) \approx s(w_{a_i}) \approx s(w_{b_j})$, which are not regarded as topic reordering. It is noteworthy to underline that Definition 3.1 only asks for an alignment if there exist two topics that share the same semantics, and if a new topic is discovered at time t_b, it is free from the requirement of being aligned.

Moreover, the last two propositions in Eq. (3.1) directly imply that the ith or the jth dimension of *document-topic* representations at time t_a and t_b represents different semantics; such a *semantic misalignment* is indicated to be the cause of a poor downstream model performance. For the convenience of describing the proposed NMTF-LTM method, we summarize the frequently used notations in Table 3.1.

NMTF-Based Objective Function A major drawback of the existing NMF-based lifelong topic model is that the discovered topics are not aligned across time, limiting the model applicability to downstream tasks. Instead of utilizing the 2-factor NMF, we exploit the 3-factor NMTF for LTM. By mapping $\mathbf{W}^{(t)}$ to an aligned topic space, the new topic representations share semantics over time. Specifically, the generated word-topic matrices $\{\mathbf{W}^{(t)}\}_{t=1}^{T}$ possess constantly changing permutations of topics; thus the semantics of each topic dimension in the corresponding document-topic matrices $\{\mathbf{H}^{(t)}\}_{t=1}^{T}$ may be misaligned. Toward addressing this issue, a linear mapping matrix $\mathbf{S}^{(t)} \in \mathbb{R}_{+}^{k \times l}$ is introduced to map the original topics into a new space, named an aligned topic space, where the permutation of topics is encouraged to be fixed and aligned. Note that l is the number of topics in the aligned space, which is considered to be consistent over time. By introducing $\mathbf{S}^{(t)}$, we can get the following NMTF-based objective function:

$$\mathcal{L}^{(t)} = \frac{1}{2} \cdot \left\| \mathbf{D}^{(t)} - \mathbf{W}^{(t)}\mathbf{S}^{(t)}\mathbf{V}^{(t)T} \right\|_F^2 + \frac{\lambda_{TM}}{2} \cdot \left\| \mathbf{D}^{(t)} - \mathbf{W}^{(t)}\mathbf{H}^{(t)T} \right\|_F^2, \tag{3.2}$$

where $\mathbf{V}^{(t)} \in \mathbb{R}_{+}^{n^{(t)} \times l}$ is the document-topic matrix based on the aligned topic space derived from $\mathbf{W}^{(t)}\mathbf{S}^{(t)}$, and λ_{TM} is the importance control parameter.

Table 3.1 The frequently used notations

Notation	Description
t	The current time
$\mathcal{D}^{(t)}$	Document chunks at time t
$\mathcal{KG}^{(t)}$	Knowledge graph at time t
$m^{(t)}$	The number of words at time t
$n^{(t)}$	The number of documents at time t
k	The number of topics in the original topic space
l	The number of topics in the aligned topic space
$\mathbf{D}^{(t)} \in \mathbb{R}_{+}^{m^{(t)} \times n^{(t)}}$	The word-document matrix of $\mathcal{D}^{(t)}$
$\mathbf{W}^{(t)} \in \mathbb{R}_{+}^{m^{(t)} \times k}$	The original word-topic matrix at time t
$\mathbf{H}^{(t)} \in \mathbb{R}_{+}^{n^{(t)} \times k}$	The original document-topic matrix at time t
$\mathbf{S}^{(t)} \in \mathbb{R}_{+}^{k \times l}$	The linear mapping between two topic spaces at time t
$\widetilde{\mathbf{W}}^{(t)} \in \mathbb{R}_{+}^{m^{(t)} \times l}$	Denotation of $\mathbf{W}^{(t)}\mathbf{S}^{(t)}$, the aligned word-topic matrix at time t
$\mathbf{V}^{(t)} \in \mathbb{R}_{+}^{n^{(t)} \times l}$	The aligned document-topic matrix at time t
$\mathbf{K}^{(t-1)} \in \mathbb{R}^{m^{(t)} \times m^{(t)}}$	The adjacency matrix of the knowledge graph involved at time t
$deg\mathbf{K}^{(t-1)} \in \mathbb{R}^{m^{(t)} \times m^{(t)}}$	The *degree* matrix of $\mathbf{K}^{(t-1)}$
$\mathbf{L}^{(t-1)} \in \mathbb{R}^{m^{(t)} \times m^{(t)}}$	The graph Laplacian matrix of $\mathbf{K}^{(t-1)}$

For clarity, we denote the product of $\mathbf{W}^{(t)}$ and $\mathbf{S}^{(t)}$ as the aligned word-topic matrix $\widetilde{\mathbf{W}}^{(t)}$, where $\widetilde{\mathbf{W}}^{(t)} \in \mathbb{R}_{+}^{m^{(t)} \times l}$. With the introduction of the aligned topic space to the original topic space, the model is now having a stirring potential of enforcing the alignment for $\widetilde{\mathbf{W}}^{(t)}$ continually with high-quality original topics in $\mathbf{W}^{(t)}$ discovered simultaneously over a text stream, in which the original topics only serve as the candidates for the aligned topics, and thus a minimal loss of topic evolution information is guaranteed for the aligned topics with adequately updated original topics. To achieve this, we further introduce two constraints in the form of regularization terms, including a consolidated regularization which aims at encouraging an alignment of semantics in the aligned topic space by reducing the change of topic permutation over time, and a regularization for KB which enables the model to retrieve knowledge from the maintained KB to enhance the original topic quality for lifelong learning. The details of both constraints will be described in the following parts.

Toward aligning semantics, we propose a **consolidated regularization term** for LTM. Inspired by knowledge distillation [77], our regularization term is defined as follows:

$$\mathcal{L}_C^{(t)}\left(\mathbf{V}^{(t)}\right) = \left\|\mathbf{D}^{(t)} - \widetilde{\mathbf{W}}^{(t-1)}\mathbf{V}^{(t)T}\right\|_F^2. \tag{3.3}$$

By introducing $\mathcal{L}_C^{(t)}\left(\mathbf{V}^{(t)}\right)$, the current topic representations of each document, i.e., the rows of $\mathbf{V}^{(t)}$, are forced to be represented by the previous aligned topics $\widetilde{\mathbf{W}}^{(t-1)}$ as well, which further leads to the alignment on the current topics $\widetilde{\mathbf{W}}^{(t)}$ in the aligned topic space. That is, $\mathbf{V}^{(t)}$ becomes the "bridge" for each topic dimension between the current topics $\widetilde{\mathbf{W}}^{(t)}$ and the previous aligned topics $\widetilde{\mathbf{W}}^{(t-1)}$. Note that the objective of this regularization is not to keep the aligned topic space absolutely consistent, but to reduce the unnecessary topic reordering and keep the aligned topic space evolving smoothly instead of changing rapidly.

Knowledge Base Following NMF-LTM [13], we maintain a KB in the form of word-word affinity graph, which is denoted as $\mathcal{KG}^{(t)}$ at time t. The nodes in $\mathcal{KG}^{(t)}$ are words, and the edge weights are word co-occurrences in topics accumulated from time 1 to t. Specifically, all word pairs among top-10 words are retrieved for every topic at the current time (i.e., the column of $\mathbf{W}^{(t)}$), and then for every word pair, we add 1 to the corresponding edge weight in $\mathcal{KG}^{(t-1)}$ and get $\mathcal{KG}^{(t)}$.

For mining the original topics $\mathbf{W}^{(t)}$ at time t, a graph regularization is employed to incorporate the previously learned knowledge in $\mathcal{KG}^{(t-1)}$:

$$\mathcal{L}_{KB}^{(t)}\left(\mathbf{W}^{(t)}\right) = \mathrm{tr}\left({\mathbf{W}^{(t)}}^T \mathbf{L}^{(t-1)} \mathbf{W}^{(t)}\right), \tag{3.4}$$

where $\mathbf{L}^{(t-1)} = deg\mathbf{K}^{(t-1)} - \mathbf{K}^{(t-1)}$ is the graph Laplacian matrix of $\mathbf{K}^{(t-1)}$, in which $\mathbf{K}^{(t-1)} \in \mathbb{R}^{m^{(t)} \times m^{(t)}}$ is the adjacency matrix constructed from $\mathcal{KG}^{(t-1)}$ using a ratio formulation [13], and $deg\mathbf{K}^{(t-1)}$ is the *degree* matrix of the graph which is computed as $diag(\mathbf{K}^{(t-1)} \cdot \mathbf{1})$.

The Overall Model By combining all the aforementioned parts, the overall objective function can be summarized as follows:

$$\begin{aligned}\mathcal{L}^{(t)} &= \frac{1}{2} \cdot \left\|\mathbf{D}^{(t)} - \mathbf{W}^{(t)}\mathbf{S}^{(t)}{\mathbf{V}^{(t)}}^T\right\|_F^2 + \frac{\lambda_{TM}}{2} \cdot \left\|\mathbf{D}^{(t)} - \mathbf{W}^{(t)}{\mathbf{H}^{(t)}}^T\right\|_F^2 \\ &+ \frac{\lambda_C}{2} \cdot \left\|\mathbf{D}^{(t)} - \widetilde{\mathbf{W}}^{(t-1)}{\mathbf{V}^{(t)}}^T\right\|_F^2 + \frac{\lambda_{KB}}{2} \cdot \mathrm{tr}\left({\mathbf{W}^{(t)}}^T \mathbf{L}^{(t-1)} \mathbf{W}^{(t)}\right),\end{aligned} \tag{3.5}$$

where λ_C and λ_{KB} are the importance control parameters. Figure 3.2 presents the framework of our NMTF-LTM which reveals the whole learning process.

The multiplicative update rules of $\mathbf{V}^{(\mathbf{t})}$, $\mathbf{W}^{(t)}$, $\mathbf{H}^{(t)}$, and $\mathbf{S}^{(t)}$ are given below:

$$\begin{aligned}\mathbf{V}^{(t)} \leftarrow \mathbf{V}^{(t)} &\odot \left({\mathbf{D}^{(t)}}^T \mathbf{W}^{(t)}\mathbf{S}^{(t)} + \lambda_C \cdot {\mathbf{D}^{(t)}}^T \widetilde{\mathbf{W}}^{(t-1)}\right) \\ &\oslash \left(\mathbf{V}^{(t)}{\mathbf{S}^{(t)}}^T{\mathbf{W}^{(t)}}^T\mathbf{W}^{(t)}\mathbf{S}^{(t)} + \lambda_C \cdot \mathbf{V}^{(t)}\widetilde{\mathbf{W}}^{(t-1)T}\widetilde{\mathbf{W}}^{(t-1)}\right),\end{aligned} \tag{3.6}$$

$$\begin{aligned}\mathbf{W}^{(t)} \leftarrow \mathbf{W}^{(t)} &\odot \left(\mathbf{D}^{(t)}\mathbf{V}^{(t)}{\mathbf{S}^{(t)}}^T + \lambda_{TM} \cdot \mathbf{D}^{(t)}\mathbf{H}^{(t)} + \lambda_{KB} \cdot \mathbf{K}^{(t-1)}\mathbf{W}^{(t)}\right) \\ &\oslash \left(\mathbf{W}^{(t)}\mathbf{S}^{(t)}{\mathbf{V}^{(t)}}^T\mathbf{V}^{(t)}{\mathbf{S}^{(t)}}^T + \lambda_{TM} \cdot \mathbf{W}^{(t)}{\mathbf{H}^{(t)}}^T\mathbf{H}^{(t)} + \lambda_{KB} \cdot deg\mathbf{K}^{(t-1)}\mathbf{W}^{(t)}\right),\end{aligned} \tag{3.7}$$

$$\mathbf{H}^{(t)} \leftarrow \mathbf{H}^{(t)} \odot {\mathbf{D}^{(t)}}^T\mathbf{W}^{(t)} \oslash \mathbf{H}^{(t)}{\mathbf{W}^{(t)}}^T\mathbf{W}^{(t)}, \tag{3.8}$$

$$\mathbf{S}^{(t)} \leftarrow \mathbf{S}^{(t)} \odot {\mathbf{W}^{(t)}}^T\mathbf{D}^{(t)}\mathbf{V}^{(t)} \oslash {\mathbf{W}^{(t)}}^T\mathbf{W}^{(t)}\mathbf{S}^{(t)}{\mathbf{V}^{(t)}}^T\mathbf{V}^{(t)}. \tag{3.9}$$

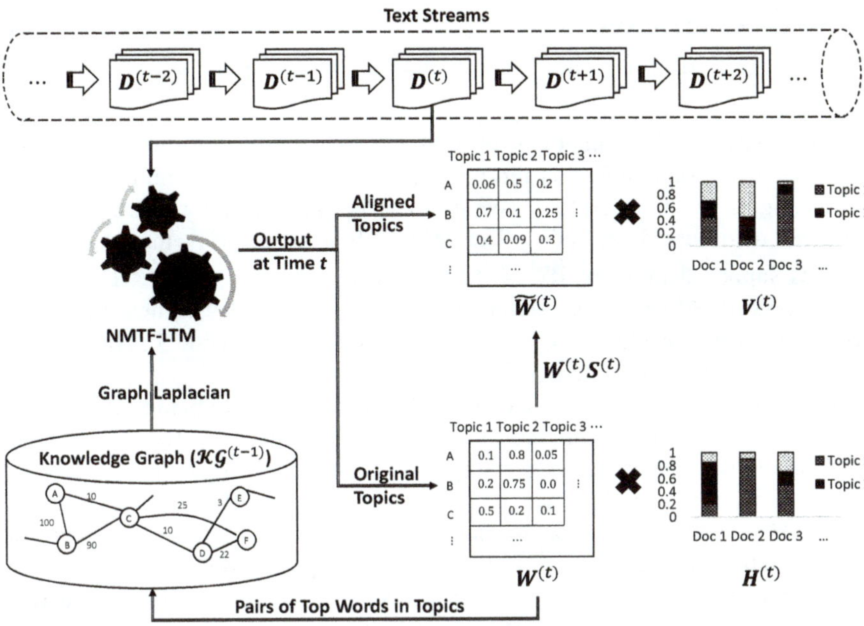

Fig. 3.2 The framework of NMTF-LTM

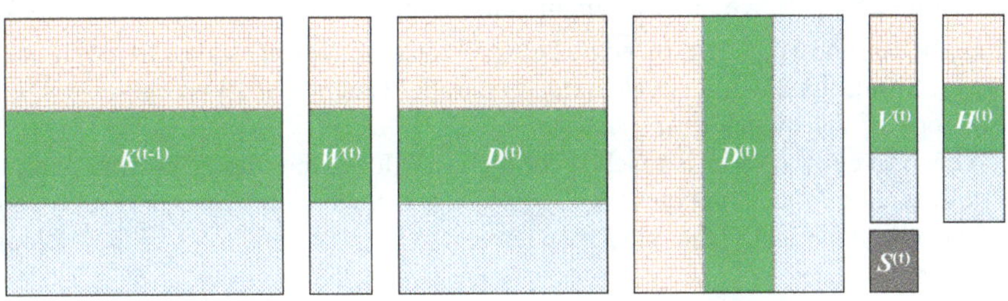

Fig. 3.3 Matrix partitioning scheme of our PNMTF-LTM

In the above, $\odot$ is the Hadamard (element-wise) product and $\oslash$ is the element-wise division for a matrix.

Parallelism for NMTF-LTM For NMTF-LTM, the multiplicative update rules suffer from intensive matrix multiplications, limiting the model scalability on large-scale data. To address this problem, we extend an existing method [12] to tackle the lifelong learning setting with extra factors and regularization terms. Specifically, we propose a data partition and distribution scheme for updating each factor and develop PNMTF-LTM by a distributed parallel update algorithm.

Figure 3.3 presents our scheme of partitioning $\boldsymbol{D}^{(t)}$, $\boldsymbol{K}^{(t)}$, and factor matrices. The partitioned matrices are of distributed storage, so that a large dataset can be stored across computing machines and can be processed in parallel. For example, each process locally stores a row block $\boldsymbol{D}_{row}^{(t)} \in \mathbb{R}^{m_p^{(t)} \times n^{(t)}}$ and a column block $\boldsymbol{D}_{col}^{(t)} \in \mathbb{R}^{m^{(t)} \times n_p^{(t)}}$ of $\boldsymbol{D}^{(t)}$, where p is the number of processes, $m_p^{(t)} \approx \frac{m^{(t)}}{p}$, and $n_p^{(t)} \approx \frac{n^{(t)}}{p}$.

After distributing those matrices, the computation can be performed by exploiting inter-process communication through MPI. The detailed algorithm is summarized in Algorithm 4, where a master process is used to perform critical serial operations, distribute data, and collect results. It is noteworthy that the master process is supposed to adapt $\widetilde{\boldsymbol{W}}^{(t-1)}$ to the vocabulary table of $\boldsymbol{D}^{(t)}$ (line 4), by deleting rows of old words and adding zero rows for new words. Then the master process constructs the adjacency matrix $\boldsymbol{K}^{(t-1)}$ from $\mathcal{KG}^{(t-1)}$ (line 5). Also, $\boldsymbol{S}^{(t)}$ is initialized by the master process (line 6).

After scattering, broadcasting, calculating, or randomly initializing block matrices (lines 8–10), the algorithm goes into the inner loop (lines 11–16), which is the key of parallelism and indicates the steps for multiplicative updating $\boldsymbol{W}^{(t)}$, $\boldsymbol{V}^{(t)}$, $\boldsymbol{H}^{(t)}$, and $\boldsymbol{S}^{(t)}$ in parallel. We take $\boldsymbol{V}^{(t)}$ as an example, and the update algorithms of $\boldsymbol{W}^{(t)}$, $\boldsymbol{H}^{(t)}$, and $\boldsymbol{S}^{(t)}$ can be similarly derived. For $\boldsymbol{V}^{(t)}$, the update rule of Eq. (3.6) can be separated into a row block-wise form and performed in a distributed framework. As shown in Algorithm 6, each process takes responsibility for updating $\boldsymbol{V}_{row}^{(t)}$ by using its local blocks $\boldsymbol{D}_{row}^{(t)}$, $\boldsymbol{W}_{row}^{(t)}$, $\widetilde{\boldsymbol{W}}_{row}^{(t)}$, and $\boldsymbol{S}^{(t)}$. Specifically, the numerator $\boldsymbol{N}_V$ of the row block-wise update rule can be *reduced* via summing up local results $\boldsymbol{N}_{V\,local}$ which are computed in each process using local matrix blocks $\boldsymbol{D}_{row}^{(t)}$, $\hat{\boldsymbol{W}}_{row}$, and $\widetilde{\boldsymbol{W}}_{row}^{(t)}$, followed by being *scattered* to each process as row blocks $\boldsymbol{N}_{V\,row}$. Similarly, $\boldsymbol{M}_{V\,row}$ is calculated in parallel except that all processes will get a full copy of $\boldsymbol{T}_V$ by using *Allreduce*.

When the convergence is reached for the inner loop, the master process gathers the results (line 17) and updates $\mathcal{KG}^{(t)}$ (lines 18–21), which finalizes the processing of data at time t. Finally, the outer loop iterates to the next time $t+1$ of the lifelong procedure.

3.1.1.3 Experimental Results

In this section, we present quantitative results of topic quality evaluation and document clustering, which indicate that our NMTF-LTM achieves a superior alignment of semantics without loss of topic quality.

Topic Quality For different models, the average C_{NPMI}, TU, and TQ scores over all chunks are summarized in Table 3.2, where "–" indicates *out of memory* for the experimental setting, and the best value in each case is marked in boldface. For each baseline, ↑ or ↓ after the result indicates that the baseline performs better or worse than our NMTF-LTM significantly according to t-tests with 5% significance level. First, we can observe that our NMTF-LTM achieves the best TQ score among datasets except for TMN. This verifies that our NMTF-LTM reaches a better

Algorithm 4: PNMTF-LTM via MPI

Input: $\{\boldsymbol{D}^{(t)}\}_{t=1}^{T}$
Output: $\{(\boldsymbol{W}^{(t)}, \boldsymbol{S}^{(t)}, \boldsymbol{V}^{(t)}, \boldsymbol{H}^{(t)})\}_{t=1}^{T}$

Let $\mathcal{KG}^{(0)} \leftarrow \emptyset$;
for $t = 1, 2, \ldots, T$ **do**
  if *master process* **then**
    Adapt $\widetilde{\boldsymbol{W}}^{(t-1)}$ to the vocabulary table of $\boldsymbol{D}^{(t)}$;
    Construct $\boldsymbol{K}^{(t-1)}$ from $\mathcal{KG}^{(t-1)}$;
    Randomly initialize $\boldsymbol{S}^{(t)}$;
  end
  Get the scattered $\boldsymbol{D}_{row}^{(t)}$, $\boldsymbol{D}_{col}^{(t)}$, $\boldsymbol{K}_{row}^{(t-1)}$, $\widetilde{\boldsymbol{W}}_{row}^{(t-1)}$, and the broadcast $\boldsymbol{S}^{(t)}$ from the *master process*;
  Calculate $deg\boldsymbol{K}_{row}^{(t-1)}$ using $\boldsymbol{K}_{row}^{(t-1)}$;
  Randomly initialize $\boldsymbol{W}_{row}^{(t)}$, $\boldsymbol{V}_{row}^{(t)}$, $\boldsymbol{H}_{row}^{(t)}$;
  repeat
    Update $\boldsymbol{W}_{row}^{(t)}$ in parallel (ref. Algorithm 5);
    Update $\boldsymbol{V}_{row}^{(t)}$ in parallel (ref. Algorithm 6);
    Update $\boldsymbol{H}_{row}^{(t)}$ in parallel (ref. Algorithm 7);
    Update $\boldsymbol{S}^{(t)}$ in parallel (ref. Algorithm 8);
  until *convergence*;
  Gather $\boldsymbol{W}^{(t)}$, $\boldsymbol{V}^{(t)}$, $\boldsymbol{H}^{(t)}$ to the *master process*;
  if *master process* **then**
    $\widetilde{\boldsymbol{W}}^{(t)} \leftarrow \boldsymbol{W}^{(t)}\boldsymbol{S}^{(t)}$;
    Update $\mathcal{KG}^{(t)}$;
  end
end

Algorithm 5: Parallel update of $\boldsymbol{W}_{row}^{(t)}$

Input: $\boldsymbol{W}_{row}^{(t)}$, $\boldsymbol{V}_{row}^{(t)}$, $\boldsymbol{H}_{row}^{(t)}$, $\boldsymbol{S}^{(t)}$, $\boldsymbol{D}_{col}^{(t)}$, $\boldsymbol{K}_{row}^{(t)}$, $deg\boldsymbol{K}_{row}^{(t)}$
Output: $\boldsymbol{W}_{row}^{(t)}$

$\hat{\boldsymbol{V}}_{row} \leftarrow \boldsymbol{V}_{row}^{(t)}\boldsymbol{S}^{(t)^T}$;
$\boldsymbol{N}_W \leftarrow \boldsymbol{D}_{col}^{(t)}(\hat{\boldsymbol{V}}_{row} + \lambda_{TM} \cdot \boldsymbol{H}_{row}^{(t)}) + \lambda_{KB} \cdot \boldsymbol{K}_{row}^{(t-1)^T}\boldsymbol{W}_{row}^{(t)}$;
$\boldsymbol{N}_{W_{row}} \leftarrow Reduce\_scatter(\boldsymbol{N}_W)$;
$\boldsymbol{T}_{W1} \leftarrow \hat{\boldsymbol{V}}_{row}^T\hat{\boldsymbol{V}}_{row} + \lambda_{TM} \cdot \boldsymbol{H}_{row}^{(t)^T}\boldsymbol{H}_{row}^{(t)}$;
$\boldsymbol{T}_{W2} \leftarrow Allreduce(\boldsymbol{T}_{W1})$;
$\boldsymbol{T}_{W3row} \leftarrow \boldsymbol{W}_{row}^{(t)}\boldsymbol{T}_{W2}$;
$\boldsymbol{T}_{W4} \leftarrow deg\boldsymbol{K}_{row}^{(t-1)^T}\boldsymbol{W}_{row}^{(t)}$;
$\boldsymbol{T}_{W4row} \leftarrow Reduce\_scatter(\boldsymbol{T}_{W4})$;
$\boldsymbol{M}_{Wrow} \leftarrow \boldsymbol{T}_{W3row} + \lambda_{KB} \cdot \boldsymbol{T}_{W4row}$;
$\boldsymbol{W}_{row}^{(t)} \leftarrow \boldsymbol{W}_{row}^{(t)} \odot (\boldsymbol{N}_{Wrow} \oslash \boldsymbol{M}_{Wrow})$;

Algorithm 6: Parallel update of $\boldsymbol{V}_{row}^{(t)}$

Input: $\boldsymbol{W}_{row}^{(t)}, \boldsymbol{V}_{row}^{(t)}, \boldsymbol{S}^{(t)}, \boldsymbol{D}_{row}^{(t)}, \widetilde{\boldsymbol{W}}_{row}^{(t-1)}$
Output: $\boldsymbol{V}_{row}^{(t)}$

$\hat{\boldsymbol{W}}_{row} \leftarrow \boldsymbol{W}_{row}^{(t)} \boldsymbol{S}^{(t)};$
$\boldsymbol{N}_{V local} \leftarrow \boldsymbol{D}_{row}^{(t)\,T} (\hat{\boldsymbol{W}}_{row} + \lambda_C \cdot \widetilde{\boldsymbol{W}}_{row}^{(t-1)});$
$\boldsymbol{N}_{V row} \leftarrow Reduce\_scatter(\boldsymbol{N}_{V local});$
$\boldsymbol{T}_{V local} \leftarrow \hat{\boldsymbol{W}}_{row}^{T} \hat{\boldsymbol{W}}_{row} + \lambda_C \cdot \widetilde{\boldsymbol{W}}_{row}^{(t-1)T} \widetilde{\boldsymbol{W}}_{row}^{(t-1)};$
$\boldsymbol{T}_{V} \leftarrow Allreduce(\boldsymbol{T}_{V local});$
$\boldsymbol{M}_{V row} = \boldsymbol{V}_{row}^{(t)} \boldsymbol{T}_{V};$
$\boldsymbol{V}_{row}^{(t)} \leftarrow \boldsymbol{V}_{row}^{(t)} \odot (\boldsymbol{N}_{V row} \oslash \boldsymbol{M}_{V row});$

Algorithm 7: Parallel update of $\boldsymbol{H}_{row}^{(t)}$

Input: $\boldsymbol{W}_{row}^{(t)}, \boldsymbol{H}_{row}^{(t)}, \boldsymbol{D}_{row}^{(t)}$
Output: $\boldsymbol{H}_{row}^{(t)}$

$\boldsymbol{N}_{H} \leftarrow \boldsymbol{D}_{row}^{(t)\,T} \boldsymbol{W}_{row}^{(t)};$
$\boldsymbol{N}_{H row} \leftarrow Reduce\_scatter(\boldsymbol{N}_{H});$
$\boldsymbol{W}^{T}\boldsymbol{W} \leftarrow Allreduce(\boldsymbol{W}_{row}^{(t)\,T} \boldsymbol{W}_{row}^{(t)});$
$\boldsymbol{M}_{H row} = \boldsymbol{H}_{row}^{(t)} (\boldsymbol{W}^{T}\boldsymbol{W});$
$\boldsymbol{H}_{row}^{(t)} \leftarrow \boldsymbol{H}_{row}^{(t)} \odot (\boldsymbol{N}_{H row} \oslash \boldsymbol{M}_{H row});$

Algorithm 8: Parallel update of $\boldsymbol{S}^{(t)}$

Input: $\boldsymbol{W}_{row}^{(t)}, \boldsymbol{V}_{row}^{(t)}, \boldsymbol{S}^{(t)}, \boldsymbol{D}_{col}^{(t)}, \boldsymbol{W}^{T}\boldsymbol{W}$
Output: $\boldsymbol{S}^{(t)}$

$\boldsymbol{T}_{S1} \leftarrow \boldsymbol{D}_{col}^{(t)} \boldsymbol{V}_{row}^{(t)};$
$\boldsymbol{T}_{S1 row} \leftarrow Reduce\_scatter(\boldsymbol{T}_{S1});$
$\boldsymbol{T}_{S2} \leftarrow \boldsymbol{W}_{row}^{(t)\,T} \boldsymbol{T}_{S1 row};$
$\boldsymbol{N}_{S} \leftarrow Allreduce(\boldsymbol{T}_{S2});$
$\boldsymbol{V}^{T}\boldsymbol{V} \leftarrow Allreduce(\boldsymbol{V}_{row}^{(t)\,T} \boldsymbol{V}_{row}^{(t)});$
$\boldsymbol{M}_{S} \leftarrow (\boldsymbol{W}^{T}\boldsymbol{W})\boldsymbol{S}(\boldsymbol{V}^{T}\boldsymbol{V});$
$\boldsymbol{S}^{(t)} \leftarrow \boldsymbol{S}^{(t)} \odot (\boldsymbol{N}_{S} \oslash \boldsymbol{M}_{S});$

comprehensive topic quality and a better balance between coherence and diversity than other baselines. To gain more insight into the performance differences of topic quality, we then illustrate the C_{NPMI} and TU results in detail. For LDA-LTM, although LDA-LTM performs well on the small TMN dataset, NMTF-LTM achieves significantly better C_{NPMI} results with competitive TU scores on the AGNews and Yahoo datasets when compared with LDA-LTM. It is also worth mentioning that LDA-LTM needs to repeatedly access the chunks in a multi-pass way. These show that LDA-LTM scales poorly to large-scale long text datasets. For LNTM, it obtains

Table 3.2 Topic quality results

Model	TMN			20News			AGNews			Yahoo		
	C_{NPMI}	TU	TQ	C_{NPMI}	TU	TQ	C_{NPMI}	TU	TQ	C_{NPMI}	TU	TQ
LDA-LTM	**0.091** ↑	**0.937** ↑	**0.085** ↑	0.064	0.772 ↓	0.049 ↓	0.081 ↓	**0.860** ↑	0.069 ↓	0.059 ↓	**0.751** ↑	0.044 ↓
LNTM	−0.388 ↓	0.713 ↓	−0.277 ↓	–	–	–	–	–	–	–	–	–
NMF-LTM	0.075	0.880 ↑	0.066	0.056 ↓	0.881 ↓	0.049 ↓	0.104	0.787	0.082	0.076	0.707 ↓	0.053 ↓
LCM	0.029 ↓	0.702 ↓	0.020 ↓	**0.137** ↑	0.506 ↓	0.058	0.053 ↓	0.658 ↓	0.035 ↓	–	–	–
NMTF-LTM	0.067	0.865	0.058	0.068	**0.896**	**0.061**	**0.108**	0.790	**0.085**	**0.077**	0.739	**0.057**

Table 3.3 Document clustering performance

	TMN		20News		AGNews		Yahoo	
Model	NMI(%)	F_β(%)	NMI(%)	F_β(%)	NMI(%)	F_β(%)	NMI(%)	F_β(%)
LDA-LTM	5.51	19.76	4.11	6.72	3.84	16.00	4.81	12.83
NMF-LTM	2.63	16.98	3.89	7.53	2.90	15.02	1.08	10.79
LCM	0.03	15.78	0.28	8.53	2.8e-3	13.36	–	–
NMTF-LTM	**22.91**	**34.21**	**22.29**	**17.62**	**21.79**	**31.04**	**16.07**	**21.16**

inferior topic quality in both coherence and diversity. When compared with NMF-LTM, our NMTF-LTM obtains a consistent topic quality verified by the fact that the differences are generally not significant statistically. For LCM, it achieves superior C_{NPMI} result with the poorest TU score on the 20News dataset, and it does not maintain a good topic quality on other datasets. Thus it is indicated to be not suitable for domain-invariant text streams. Based on the analysis above, it can be concluded that our NMTF-LTM achieves a promising topic quality when compared with the baselines of lifelong topic models.

Alignment Quality The evaluation results of alignment quality, i.e., document clustering results, are shown in Table 3.3, where "–" indicates *out of memory* and the best values are marked in boldface. According to t-test results with 5% significance level, the superiorities of our NMTF-LTM are statistically significant for all metrics. We do not include the results of LNTM since its document representation cannot be interpreted as a topic distribution. It can be observed that our NMTF-LTM achieves remarkable document clustering performances, which significantly outperforms these baselines on all datasets by a large margin. Besides, the performance differences between our NMTF-LTM and each baseline are statistically significant by performing t-tests on all metrics. This evidently shows the effectiveness of our NMTF-LTM on generating an aligned topic space.

Topic-Level Dynamic Information Retrieval The P@10 performances of topic-level dynamic IR are depicted in the form of heatmap, as shown in Fig. 3.4, where the horizontal axis indicates the current chunk id, and the vertical axis indicates the previous chunk id. Since it is too space-consuming to perform topic-level dynamic IR on the Yahoo dataset for all baselines and the latent representations generated by LNTM are not topic distributions, the corresponding results are not included in this task. Evidently, when the current topic representations are not aligned with previous topic representations, even if those documents in previous chunks share similar semantics with the given query, they cannot be accurately retrieved, resulting in a low P@10 result. Thus, a higher P@10 indicates a better alignment quality, and it can be clearly observed that our NMTF-LTM achieves the best topic-level dynamic IR performances as compared with other baselines.

Online Document Classification Under the framework of a topic-guided deep neural network [29] with the *document-topic* representations of each chunk as the

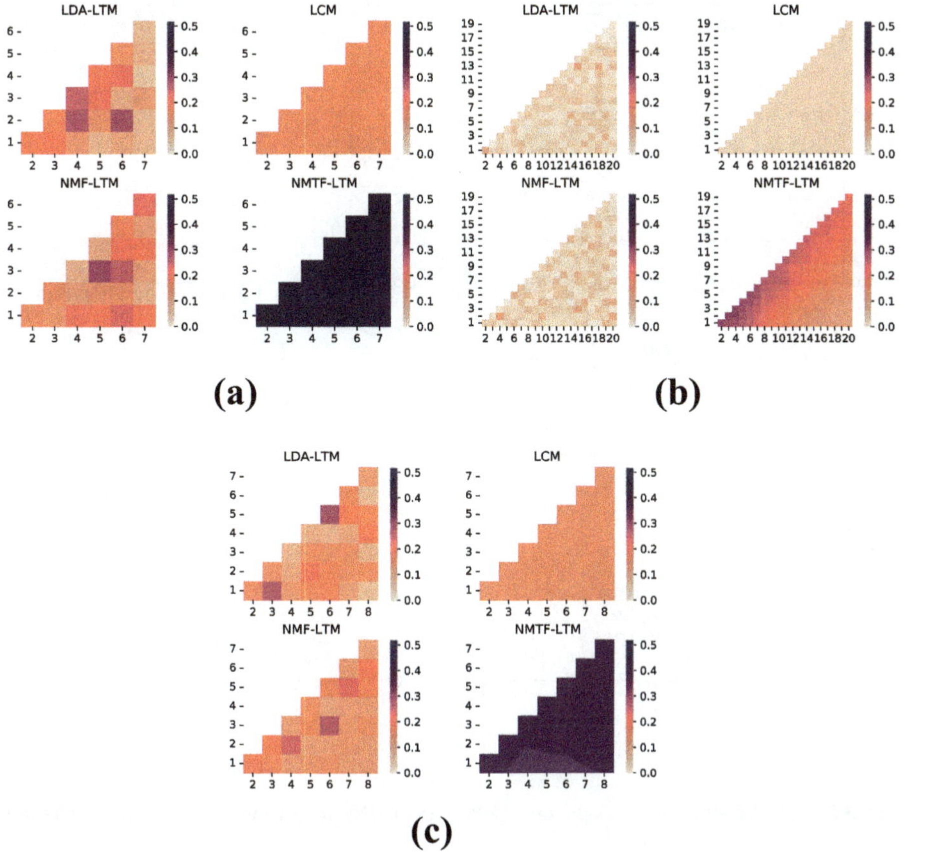

Fig. 3.4 Topic-level dynamic IR performance (P@10). (**a**) TMN. (**b**) 20News. (**c**) AGNews

input mini-batches, the online classification accuracy scores over time are plotted in Fig. 3.5. This task is not conducted for LNTM due to a lack of generating topic representations. The results indicate that our NMTF-LTM generally achieves the best accuracy scores among all datasets. We can also observe that the accuracy scores of baselines decline over time and the curves are shaking for most cases due to the rapid semantic drift caused by topic reordering, while our NMTF-LTM always enjoys valid online learning procedures with stable or rising trends of accuracy scores. It is worth noting that although there are some outliers for NMTF-LTM in Fig. 3.5b since the number of documents per chunk is relatively small which brings unstable training, our NMTF-LTM keeps dominating other baselines and the overall accuracy trend is still rising. In conclusion, our NMTF-LTM can achieve a good semantic alignment quality, and such an alignment is beneficial to the downstream task of online document classification.

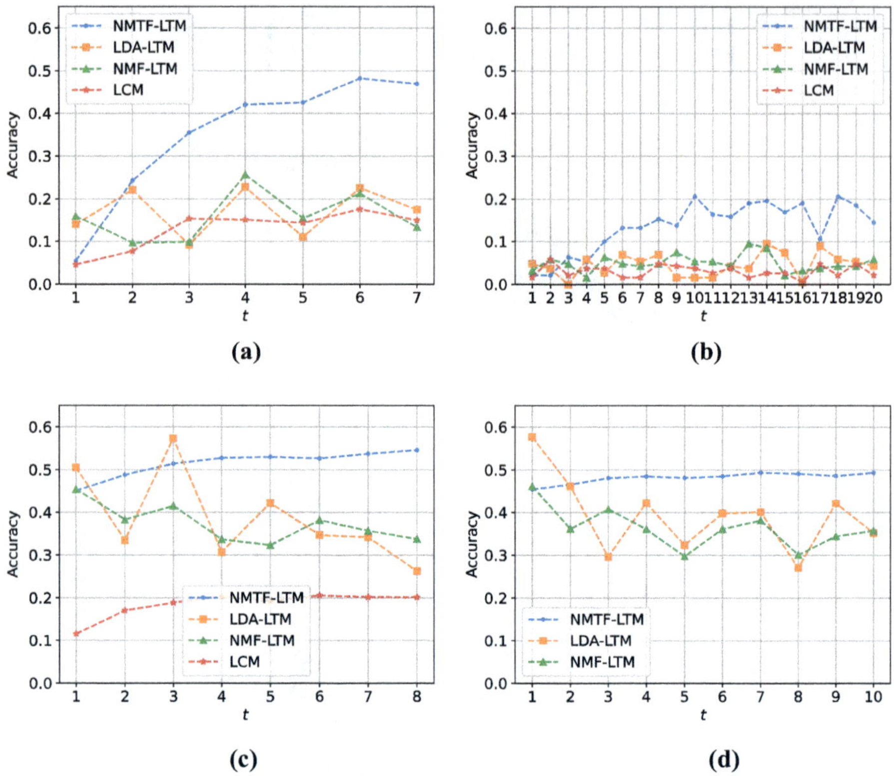

Fig. 3.5 Online document classification accuracy. (**a**) TMN. (**b**) 20News. (**c**) AGNews. (**d**) Yahoo

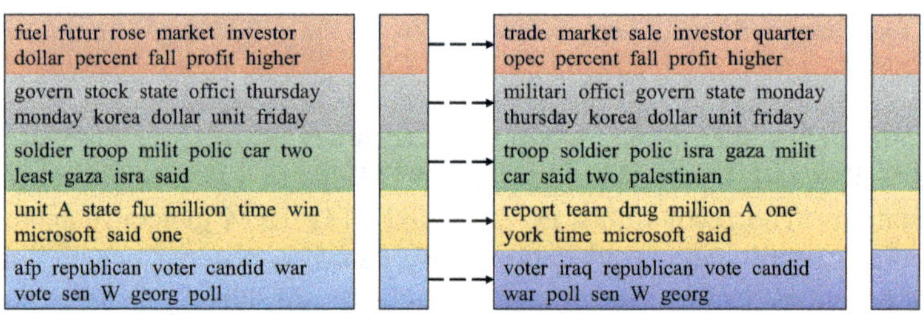

Fig. 3.6 Semantic alignment of NMTF-LTM on AGNews

Qualitative Analysis Recall that Fig. 3.1 presents a semantic misalignment example obtained by NMF-LTM with five topics on the AGNews dataset, where the topic reordering happens. On the contrary, it can be observed from Fig. 3.6 that our NMTF-LTM manifests distinguishing advantages in the following three aspects. Firstly, there are no changes in the order of topics, indicating that our NMTF-LTM achieves an alignment of semantics successfully. Secondly, even with only five topics, we can clearly identify the semantics of each topic as *Business*, *Finance*, *War*, *USA*, and *Politics*, respectively. This reveals nice interpretability and quality of the discovered topics. Thirdly, since the semantics of topics are aligned, we can easily track the evolution of topic semantics. For example, the word "palestinian" was newly discovered in the topic of *War*, revealing the evolving pattern of this topic. This also validates that our NMTF-LTM can simultaneously achieve semantic alignment and new topic discovery.

To further illustrate the capability of our NMTF-LTM in capturing the evolution of topics, we present an example of the 12th aligned topic on the 20News dataset whose number of chunks is the largest so that the evolution can be sufficiently revealed. Note that different from the baseline models, the evolution of a specific topic can be conveniently tracked for our NMTF-LTM owing to an alignment of semantics. As shown in Table 3.4, the 12th topic represents the semantics of *Urban* in the beginning, while it turns into an intermediate topic of *Sports* at the third time, which is derived from the eighth topic as shown in Fig. 3.7. This also marks the sudden disappearance of the *Urban* topic by our overall observation. Then, the 12th topic merges into the eighth topic again, and it finally emerges as a new topic of *Vehicle* at the fourth time which is also a new topic discovered in the original topic space. Note that although the semantics of this topic is retained, the representative

Table 3.4 Top 10 words of the 12th topic generated by NMTF-LTM over 20 document chunks of 20News

t	Top words
1	oakland tiger boston francisco detroit new nation citi ranger york
2	St loui dodger brave cincinnati giant chicago minnesota score york
3	get streaki walk behind punjabi runner pitcher stanford leland ball
4	GT like first mile wave oil sell cc model blue
5	honda get new buy good road may dod counterst dealer
6	motorcycl new front turn light dog small auto safeti ride
7	ride V current good tire bike releas year clutch class
8	probe sho power It rear mile new saturn ford profit
9	good year alarm design rear probe buy insur light saturn
10	mph odomet small the owner honda driver speed back civic
11	get ride quick behind saturn hand engin dod front honda
12	one insur engin MR dod rider wheel road passeng front
13	pay helmet sell blue rider year ford the K mile
14	vehicl passeng volvo wax wheel auto year bmw price ride
15	door good automat speed com they bike sport liter insur
16	US owner road look good manual fluid mile tire gear
17	speed drive turn chang mile clutch K mileag transmiss callison
18	It US road clutch speed owner seat mile power integra
19	dod offer V bought price sell mileag junk odomet rider
20	honda brake lock dealer buy one year mile convert dog

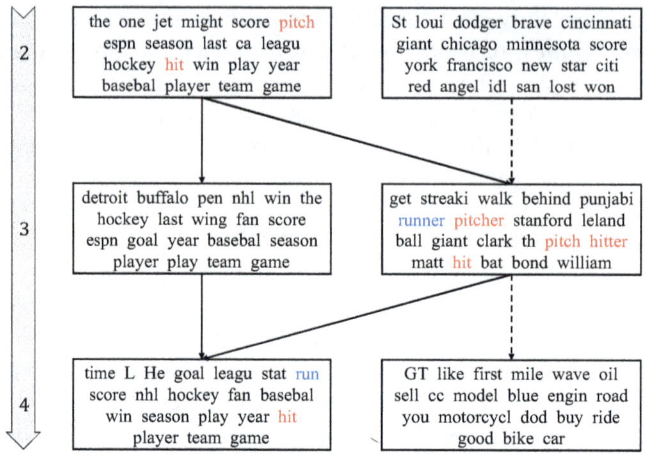

Fig. 3.7 Evolution of topics 8 and 12 from NMTF-LTM across 3 document chunks in 20News

topic words still keep evolving over time, revealing the change of focus for this topic. In conclusion, our NMTF-LTM that introduces the aligned topic space is capable of tracking the evolution of topics, such as the emergence of new topics, the splitting of one topic into two topics, the merging of two topics into one topic, and the sudden disappearance of a topic.

3.1.2 Lifelong Hierarchical Topic Model

In this section, we describe a Lifelong Nonparametric Clustering-based Hierarchical Topic Model (LNCHTM). A nonparametric clustering method offers flexibility to the topic hierarchy generation by automatically determining the number of clusters to be generated. To automatically accumulate information of hierarchical structures, we design a knowledge base in the form of a word hierarchy. To utilize it, we adopt heuristic guidance in Affinity Propagation (AP) using the similarity between words from the corpus and words in the knowledge base. The update of the knowledge base can be viewed as a completion of a partial tree. To generate more coherent and diverse topics, we add co-occurrence graph guidance on word embeddings and representative word filtering during the top-down clustering process.

3.1.2.1 Introduction

Most existing Hierarchical Topic Models (HTMs) hold a strict assumption of the topic structure, such as the fixed depth and breadth. In this case, their generalization

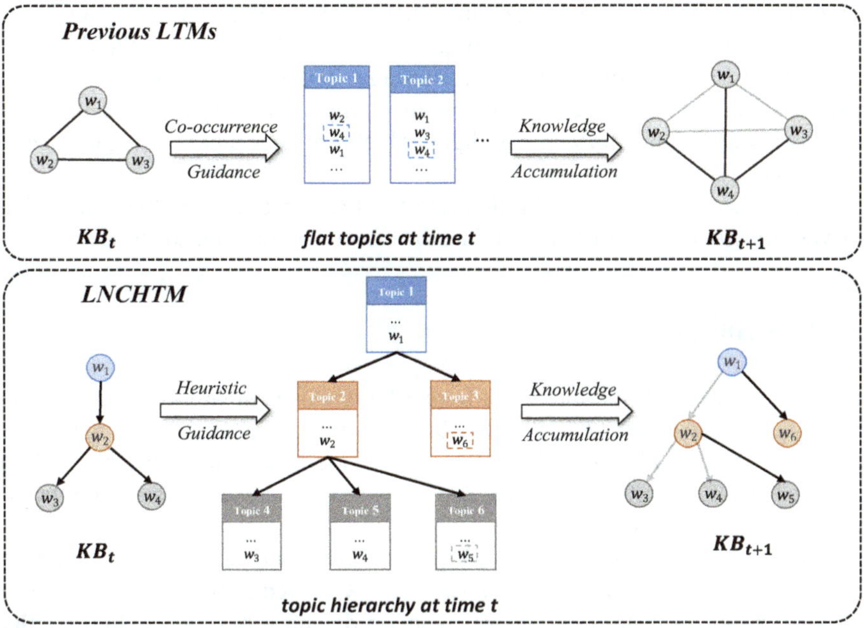

Fig. 3.8 Comparison on exploiting and updating the knowledge bases of flat LTMs (i.e., NMF-LTM [13] and NMTF-LTM [38]) and LNCHTM

ability is limited when confronted with the varied distributions of a document stream. Moreover, it has been proved that the prior knowledge of hierarchical structure can help form a more rational topic hierarchy [36, 37, 52, 78]. Unfortunately, the manual work on acquiring the prior knowledge of hierarchical structure is hard and expensive [67]. A promising solution is to automatically accumulate knowledge from the past into a knowledge base by using the lifelong learning paradigm. Currently, there have been some Lifelong Topic Models (LTMs) on document streams [13, 16, 28, 38, 58]. Nevertheless, most existing LTMs are flat topic models, and their knowledge bases only store the word relations within one topic or the distinct topic-word representations.

To address the issues mentioned above, we propose an LNCHTM over document streams. Notably, we adopt the AP algorithm [25] to automatically determine the number of topics in different levels by grouping word embeddings, which enables our model to learn a flexible topic hierarchy. As shown in Fig. 3.8, while previous LTMs (i.e., NMF-LTM [13] and NMTF-LTM [38]) only store symmetric correlations among words, we build a knowledge base in the form of word hierarchy to store the asymmetric hierarchical relation information from the past. As the stream of documents continuously emerges, the hierarchy in the knowledge base can be expanded to store more structural information. In summary, our main contributions are as follows:

- We are the first to conduct nonparametric word embedding clustering for hierarchical topic modeling, which can automatically determine the depth and breadth of topic hierarchy, therefore forming a flexible topic structure.
- We construct a knowledge base of word hierarchy to store asymmetric knowledge following a lifelong learning paradigm, which provides a heuristic guidance for topic hierarchy construction.
- We are the first to highlight the catastrophic forgetting problem in LTMs. Experiments on three metrics validate the effectiveness of our method.

3.1.2.2 Methodology

Here we formalize the problem of lifelong hierarchical topic modeling. Given n document chunks $\mathcal{D} = \{D_1, D_2, \ldots, D_n\}$, we aim to generate topic hierarchies $\mathcal{H} = \{H_1, H_2, \ldots, H_n\}$ with the aid of knowledge base. Each topic hierarchy is a directed acyclic graph, where each node is a topic and each edge directs from a parent topic to one of its child topics.

To automatically construct a prior hierarchy and improve the model flexibility, we propose the LNCHTM, which consists of a **Knowledge Base (KB)** for storing the extracted hierarchical relations of topic words and a KB-guided **hierarchical word embedding clustering** module for topic hierarchy construction. Specifically, LNCHTM regards each cluster of words as a topic and constructs the parent-child relationship among topics naturally via the top-down clustering procedure. Besides, our KB is a hierarchy of representative words, which provides heuristic guidance by aligning the newly formed topic structure with the stored structure and can be updated using the newly formed hierarchy automatically. To generate coherent and diverse topics, we also introduce **co-occurrence graph guidance** and **representative word filtering** in hierarchical word embedding clustering. The framework of our LNCHTM is shown in Fig. 3.9. The detailed steps of our LNCHTM are given in Algorithm 9.

Nonparametric Word Clustering Inspired by the parent-child relationship in the hierarchical structure, we perform top-down hierarchical clustering on word embeddings to generate topic clusters level by level, which imitates a natural derivation process. For the chunk $D_t \in \mathcal{D}$, words in its vocabulary $\mathcal{V}_t$ are first divided into several clusters/topics $C_t^{(1)} = \{C_{t,1}^{(1)}, C_{t,2}^{(1)}, \ldots\}$ in level-1, where each cluster $C_{t,i}^{(1)} \in C_t^{(1)}$ contains a set of words $\mathcal{V}_{t,i}^{(1)} \subset \mathcal{V}_t$. Then for a coarse topic $C_{t,i}^{(1)}$, we further conduct clustering on $\mathcal{V}_{t,i}^{(1)}$ to obtain its children, i.e., $C_{t,i}^{(2)} = \{C_{t,i,1}^{(2)}, C_{t,i,2}^{(2)}, \ldots\}$ in level-2. The above clustering process repeats until the entire topic hierarchy is constructed. The number of children is automatically determined by the nonparametric clustering algorithm below, and we only filter those clusters whose size is too small (i.e., smaller than 15 because we need to take top-15 words for topic evaluations).

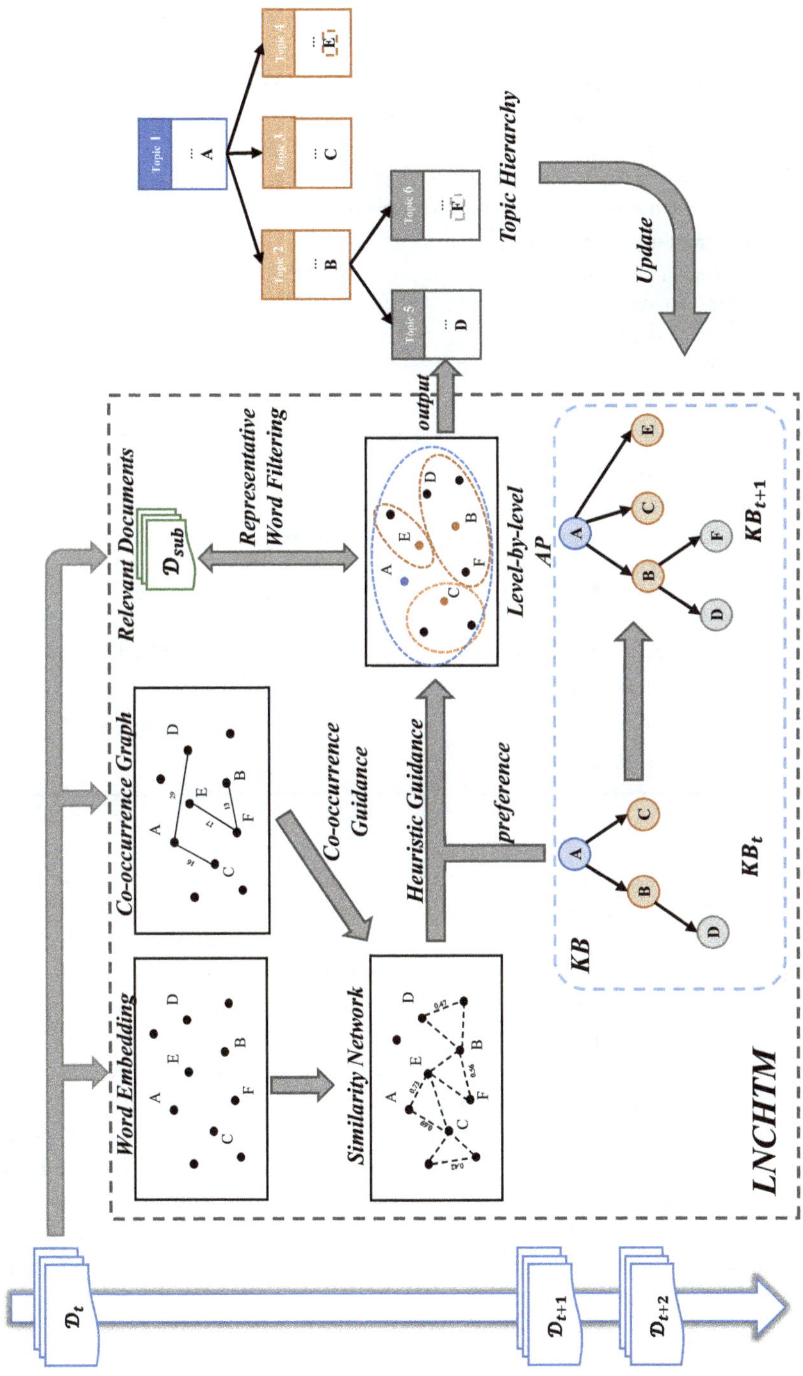

Fig. 3.9 The framework of LNCHTM

Algorithm 9: The workflow of our LNCHTM

Input: document chunks $\mathcal{D} = \{D_1, D_2, \ldots, D_n\}$, pretrained word embeddings $\boldsymbol{E}$ (e.g., GloVe);

Output: Topic hierarchies $\mathcal{H} = \{H_1, H_2, \ldots, H_n\}$, knowledge base $\mathcal{K}_n$;

Test on a random chunk to obtain δ and initialize $\mathcal{K}_0$;
for $t = 1, 2, \ldots, n$ **do**
  Obtain the vocabulary $\mathcal{V}_t$ and word embeddings $\boldsymbol{E}_t$ of D_t;
  Initialize H_t with a root, corresponding to cluster $C_t^{(0)} = \boldsymbol{E}_t$;
  Assign the root node of $\mathcal{K}_{t-1}$ to $C_t^{(0)}$;
  Initialize a queue $Q = \{C_t^{(0)}\}$;
  while *Q is not empty* **do**
    $C_t^{(j)} \leftarrow Q.pop()$;
    Obtain the vocabulary $\mathcal{V}_t^{(j)}$, word embeddings $\boldsymbol{E}_t^{(j)}$, and co-occurrence graph $\boldsymbol{C}_t^{(j)}$ for $C_t^{(j)}$;
    Obtain enhanced embeddings $\widehat{\boldsymbol{E}}_t^{(j)}$ via the guidance of $\boldsymbol{C}_t^{(j)}$ (Eq. (3.12));
    Calculate preference p if $C_t^{(j)}$ is guided;
    Calculate the similarity matrix $\boldsymbol{S}_t^{(j)}$ with p and $\boldsymbol{E}_t^{(j)}$;
    Conduct AP algorithm on $\boldsymbol{S}_t^{(j)}$ to obtain the clusters $C_{t,1}^{(j+1)}, C_{t,2}^{(j+1)}, \ldots, C_{t,m}^{(j+1)}$; ($m$ is determined automatically by AP)
    Extract topics $T_0, T_1, \ldots, T_m$ with the top-K center words;
    Fit topics $T_0, T_1, \ldots, T_m$ into H_t;
    Assign the child nodes in $\mathcal{K}_{t-1}$ to $C_{t,1}^{(j+1)}, C_{t,2}^{(j+1)}, \ldots, C_{t,m}^{(j+1)}$ is guided;
    Filter representative words from $C_{t,1}^{(j+1)}, C_{t,2}^{(j+1)}, \ldots, C_{t,m}^{(j+1)}$;
    Add $C_{t,1}^{(j+1)}, C_{t,2}^{(j+1)}, \ldots, C_{t,m}^{(j+1)}$ to Q if their sizes are valid for clustering;
  Update $\mathcal{K}_{t-1}$ with H_t to obtain $\mathcal{K}_t$;

For the choice of the clustering algorithm, we adopt the AP algorithm [25] that can automatically determine the number of clusters, so the hierarchical structure can be more flexible. For example, the number of one cluster's children can be different from its siblings, and the depth of each sub-tree can be different. Oppositely, most of the prior hierarchical topic models require predefined numbers of topics in each layer, which is hard for human decisions.

To meticulously explain a single-word clustering process for subtopic discovery, let us take a coarse topic C that contains a set of words $\mathcal{V}_C$ ($N = |\mathcal{V}_C|$) as an example. AP regards the data points as a network and passes real-valued messages through it to eventually choose a set of good exemplars and form the corresponding clusters. According to AP, we construct a similarity network $\mathbf{S} \in \mathbb{R}^{N \times N}$ among words, where $\mathbf{S}_{i,j}$ denotes the cosine similarity between word i and j in the embedding space.[1] Then we discover word clusters via the message

[1] In this work, we employ pretrained GloVe word embeddings downloaded from https://nlp.stanford.edu/projects/glove. It is also feasible to opt for other pretrained embeddings (e.g., those of BERT or RoBERTa) or to train embeddings from scratch on the target corpus.

passing procedure in the similarity network. There are two kinds of message: (1) the Responsibility matrix $\mathbf{R} \in \mathbb{R}^{N\times N}$ where $\mathbf{R}_{i,j}$ represents how suitable word j is to be the exemplar for i and (2) the Availability matrix $\mathbf{A} \in \mathbb{R}^{N\times N}$ where $\mathbf{A}_{i,j}$ reflects the appropriateness for word i to choose j as its exemplar. Formally, $\mathbf{R}$ and $\mathbf{A}$ are calculated as follows:

$$\mathbf{R}_{i,j} = \begin{cases} \mathbf{S}_{i,j} - \max\limits_{k\neq j}\{\mathbf{S}_{i,k}\}, & i = j \\ \mathbf{S}_{i,j} - \max\limits_{k\neq j}\{\mathbf{A}_{i,k} + \mathbf{R}_{i,k}\}, & i \neq j \end{cases} \tag{3.10}$$

$$\mathbf{A}_{i,j} = \begin{cases} \sum\limits_{k\neq j} \max\{\mathbf{R}_{k,i}, 0\}, & i = j \\ \min\{0, \mathbf{R}_{j,j} + \sum\limits_{k\neq i,j} \max\{\mathbf{R}_{k,j}, 0\}\}, & i \neq j. \end{cases} \tag{3.11}$$

Next, exemplars for clusters are identified based on the responsibility and availability matrices. Let $e_i = \arg\max_k\{\mathbf{R}_{i,k} + \mathbf{A}_{i,k}\}$. If $e_i = i$, then word i is an exemplar; otherwise e_i is the exemplar for word i. The convergence reaches when the changes of responsibility and availability matrices are below a manual threshold or the exemplar decisions are stable for some iterations.

Later we introduce the **co-occurrence graph guidance** and **representative word filtering** algorithms, which can effectively improve the quality of mined topics.

Co-occurrence Graph Guidance A set of facts is said to be coherent if there is a context covering most of the facts [60]. Therefore, a topic is more likely to be coherent if the word pairs in the topic occur more in the same document. Based on this insight, we enhance the input with word co-occurrence graph guidance in each document chunk. Following Liu et al. [45], we adopt a transformation on word embeddings to improve the similarities between word pairs with high co-occurrence:

$$\widehat{\mathbf{C}} = \lambda\mathbf{C} + \mathbf{I}, \quad \widehat{\mathbf{D}} = deg(\widehat{\mathbf{C}})$$
$$\widehat{\mathbf{E}} = \widehat{\mathbf{D}}^{-1}\widehat{\mathbf{C}}\mathbf{E}, \tag{3.12}$$

where $\mathbf{C}$ denotes the co-occurrence matrix and $\mathbf{E}$ denotes the original embedding matrix. For lower-level clustering, the co-occurrence matrix can be extracted from the original co-occurrence matrix.

Representative Word Filtering Notably, some words should remain in the parent topic instead of being allocated to any child topics [78]. In this case, we attempt to measure the representativeness of a word corresponding to a topic by the BM25 relevance. Firstly we extract documents in the document chunk which are relevant to the current topic by selecting those containing the top-N topic words. Then we can calculate the BM25 relevance of a word by treating it as a single-word query. For a query-like BM25 calculation, $Score(q, d_j) = \sum_{w_i \in q} W_i \cdot Rd(w_i, d_j) \cdot Rq(w_i, q)$, which can be viewed as a variant of TF-IDF. The IDF term is adjusted to $W_i = \log(\frac{N - df_i + 0.5}{df_i + 0.5})$, and the TF term is weighted like $Rd(w_i, d_j) = \frac{(k_1+1) f(w_i, d_j)}{f(w_i, d_j) + k_1(1 - b + b\frac{l_{d_j}}{l_{avg}})}$. Moreover, an intra-query weight is added: $Rq(w_i, q) = \frac{(k_2+1) f(w_i, q)}{k_2 + f(w_i, q)}$, where k_1, k_2, b are hyperparameters, df_i is the number of documents containing w_i, $f(w_i, d_j)$ is the count of w_i in d_j, and $f(w_i, q)$ is the count of w_i in q.

Knowledge Base To store knowledge of the hierarchical topic structure and topic semantics, we develop a KB in a tree structure that is consistent to the topic hierarchy. Each node in KB is one representative word. On the one hand, our KB serves as a partial topic hierarchy to guide the clustering procedure. On the other hand, it is updated by the constructed topic hierarchies continually. Formally, our KB is denoted as $\mathcal{K} = (\mathcal{N}, \mathcal{E})$, where $\mathcal{N}$ is the set of nodes (i.e., representative words) and $\mathcal{E}$ contains hierarchical relations among nodes.

Utilization of KB To align the current topic structure with the partial topic structure in KB, we design a heuristic AP clustering paradigm. Our key idea is to estimate the possibility of each word to become the cluster center (i.e., preference) based on KB. First, for each word w_i, we find the node n_i in KB that is the most similar one to it, i.e., $n_i = \arg\max_{n \in \mathcal{N}} sim(w_i, n)$. Then we adopt the maximum similarity between w_i and children of n_i as the preference of w_i, which is defined as follows:

$$p_i = \max_{n_j \in children(n_i)} \{sim(w_i, n_j)\} \cdot \min(\mathbf{S}). \tag{3.13}$$

Notably, the diagonal elements $\{\mathbf{S}_{i,i}\}$ of the similarity matrix are set to $\{p_i\}$, which guides the message passing process in AP [25]. Thereby words with higher values of preference are more likely to be the exemplars.

Maintenance of KB To match the generated topic clusters with children nodes in KB, we calculate the similarities between exemplars and child nodes. Then we allocate the cluster with maximum similarity to be the guided cluster of the corresponding child node, while other clusters are unguided:

$$guided\ index_j = \arg\max_i \{sim(e_i, n_j)\}, \tag{3.14}$$

where e_i is the exemplar for topic i and n_j is node j in KB.

With such a match, we can maintain our KB with a new topic structure by:

- *Node Attachment*: For an unguided cluster with a guided parent, if the quality of the topic is above a threshold δ, a new node can be inserted into KB with the exemplar being the representative word.

- *Word Substitution*: For a guided cluster, if the quality of the topic is above δ, the representative word can be replaced by the exemplar.
- *Node Removal*: For a guided cluster, if the quality of the topic is below δ, we can add a count for it and minus a count conversely. If the count for a node reaches a threshold σ, the node is removed from KB with all descendants lost. The count threshold is related to the number of chunks, which can be interpreted as times of failure guidance.

3.1.2.3 Experimental Results

We conduct our experiments on three varied benchmark datasets. 20News consists of over 18,000 documents of news involving 20 categories. IMDB collected by Maas et al. [47] includes 50,000 movie reviews from IMDB with label 0 marking the rating of the review is below 5 and label 1 representing the rating is above 7. Grolier is sampled from Grolier Multimedia Encyclopedia, which contains 30,024 documents covering various fields.

To evaluate the quality of topics, we take two widely used metrics, i.e., topic coherence C_{npmi} [60] and topic uniqueness (TU) [55]. Following Dieng et al. [20], TQ is calculated by multiplying C_{npmi} and TU to represent the integrated topic quality. The quality scores are calculated by the average score of top-$N \in [5, 10, 15]$. According to the empirical results, it is recommended that the value of ϵ in the definition of NPMI is set to 0 and the undefined calculation (i.e., when $P(w_i, w_j) = 0$) is default to 0 [42].

For the hierarchical structure, we focus on the relations between parents and children. In this case, Chen et al. [15] proposed the Cross-Level NPMI (CLNPMI) to quantify the coherence and the Overlap Rate (OR) can be applied for redundancy evaluation.

As a lifelong learning paradigm, we also pay attention to the catastrophic forgetting problem. The average score (A_t): $A_t = \frac{1}{t}\sum_{r=1}^{t} S_{t,r}$, the forgetting score ($F_t$): $F_t = \frac{1}{t-1}\sum_{r=1}^{t-1} \max_{r' \in [1,t-1]}(S_{r',r} - S_{t,r})$, and the learning score (LA_t): $LA_t = \frac{1}{t}\sum_{r=1}^{t} S_{r,r}$ are proposed for testing catastrophic forgetting [49], where $S_{t,r}$ denotes the score on task r after training on task t. The average score indicates how effective the knowledge base of time t is for the overall task by adopting the knowledge base on every chunk and averaging the scores. The forgetting score reflects the average decline of performance after learning chunk t compared to the best performance before it on the overall task. The learning score represents the average score applying the knowledge base t directly to the task t, which somewhat embodies the direct learning effect. These three evaluation metrics offer different perspectives on how the LTM suffers from catastrophic forgetting.

The average C_{npmi} and TU scores are summarized in Table 3.5. Obviously, LNCHTM achieved the best topic coherence on all datasets and is obviously superior to most of the baselines except NMTF-LTM. Compared with NMTF-LTM, our TU is much better on all datasets by an average margin of 0.116 on 20News,

Table 3.5 Model performance on topic quality. * denotes that the model is nonparametric, so no fixed topic number can be listed

Model	N_{topics}	20News			Grolier			IMDB		
		TC	TU	TQ	TC	TU	TQ	TC	TU	TQ
ECRTM	50	0.289	0.960	0.277	0.232	0.922	0.214	0.126	**0.968**	0.122
	200	0.115	**0.981**	0.112	0.155	**0.983**	0.153	0.061	0.953	0.058
NMF-LTM	50	0.268	0.898	0.241	0.143	0.124	0.018	0.060	0.143	0.008
	200	0.233	0.746	0.176	0.143	0.037	0.005	0.061	0.046	0.003
NMTF-LTM	50	0.346	0.877	0.303	0.357	0.885	0.316	0.127	0.725	0.092
	200	0.329	0.801	0.264	0.311	0.794	0.247	0.132	0.622	0.082
TaxoGen	*	0.297	0.890	0.264	0.187	0.920	0.172	0.126	0.884	0.111
CluHTM	*	0.285	0.276	0.079	0.304	0.350	0.106	0.108	0.203	0.022
nTSNTM	*	0.220	0.712	0.157	0.317	0.710	0.225	0.117	0.625	0.073
NSEM-GMHTM	8/32/128	0.256	0.804	0.206	0.334	0.770	0.257	0.119	0.551	0.065
LNCHTM	*	**0.372**	0.955	**0.355**	**0.373**	0.958	**0.358**	**0.155**	0.948	**0.147**

Table 3.6 Model performance on hierarchical topic quality

Dataset	Metric	TaxoGen	CluHTM	nTSNTM	NSEM-GMHTM	LNCHTM
20News	CLNPMI	0.182	0.189	0.116	0.121	**0.235**
	OR	0.106	0.260	0.091	**0.013**	0.045
Grolier	CLNPMI	0.151	0.200	0.201	0.198	**0.276**
	OR	0.083	0.163	0.104	**0.018**	0.042
IMDB	CLNPMI	0.080	0.062	0.070	0.058	**0.089**
	OR	0.113	0.101	0.137	**0.041**	0.050

0.118 on Grolier, and 0.274 on IMDB. When it comes to ECRTM, which achieved the best topic uniqueness, the average gap between LNCHTM and ECRTM is tiny with 0.015 on 20News, − 0.055 on Grolier, and 0.0125 on IMDB, while the gap of topic coherence is much bigger. From the evaluation on flat topic quality, we can conclude that LNCHTM generates comprehensively optimal topics.

In the hierarchical evaluation, we applied the original setting of constructing edges between parents and children. For example, the neural hierarchical topic models take the 2 most relevant child nodes as children [23], and the clustering-based methods (i.e., TaxoGen and LNCHTM) take every child formed by clustering as children. The results are listed in Table 3.6. On all datasets, we obtained the best CLNPMI, especially on 20News and Grolier. As for the redundancy measure, our method is slightly inferior to NSEM-GMHTM, while it is still better than others and the values of LNCHTM are below 0.05, which are acceptable for redundancy. It is worth mentioning that parent and child topics need to share some words to be coherent with each other, so a lower OR is not always better [24]. Considering that there are quite more pairs of parent-child taken by LNCHTM than NSEM-GMHTM, our representativeness measure was proved useful for redundancy reduction.

For comprehensive comparison with both flat and hierarchical LTMs, Fig. 3.10 illustrates the average score, forgetting score, and learning score of C_{npmi} changing with chunks. For each dataset (i.e., each row), the subfigures from left to right represent the results of average score (as), forgetting score (fs), and learning score (ls), respectively. Notably, LNCHTM achieved the best performance on average score and learning score over all time on the three datasets. It reveals that LNCHTM has a good learning ability to continually generate topics with high quality and the knowledge base stores general information that is suitable for different chunks. As for the forgetting score, LNCHTM achieved a competitive score compared with other LTMs on 20News.

To intuitively illustrate the flexibility of LNCHTM in hierarchical topic structure learning, we visualize a partial topic hierarchy generated by LNCHTM on chunk 5 of IMDB, as shown in Fig. 3.11. We can observe that each topic at level-1 has different numbers of children. For the second topic, the content is about the air crash, and its two subtopics are around the mechanical components of an airplane, in addition to the repair and detection. The third topic is related to guns, with child topics of the gun policy, the assassins, the cowboy (western character who usually

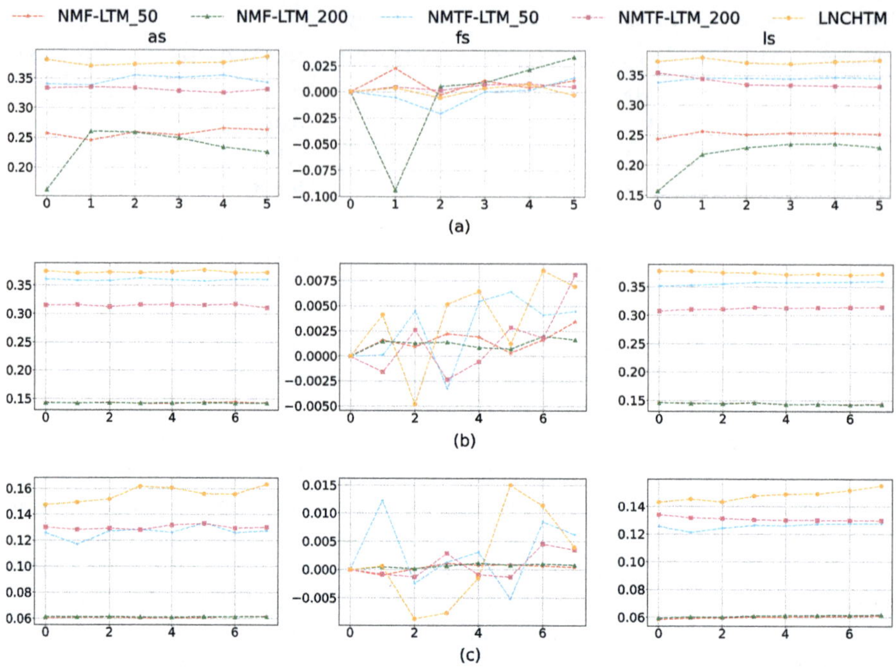

Fig. 3.10 Model performance comparison on catastrophic forgetting. (**a**) 20News. (**b**) Grolier. (**c**) IMDB

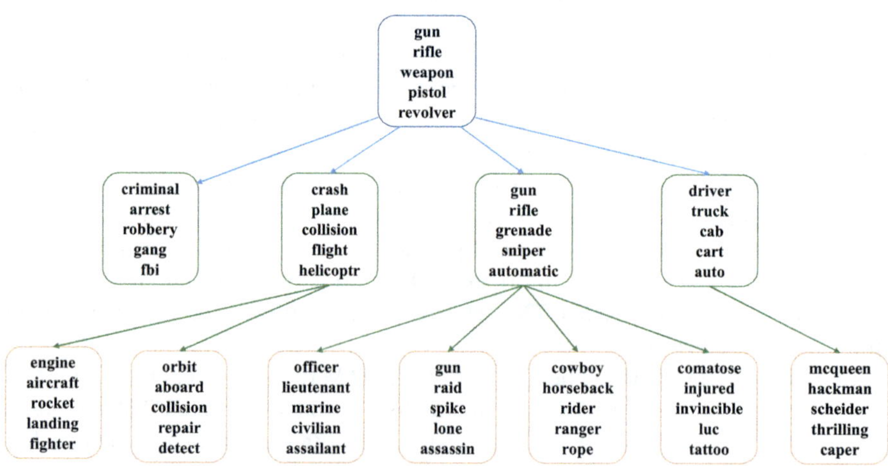

Fig. 3.11 Partial topic hierarchy on IMDB by LNCHTM

wields guns), and the injury. The only child of the fourth topic, which talks about the different categories of cars, is consistent with its parent. In conclusion, the example is a coherent topic hierarchy with flexibility in breadth.

3.2 Embedding-Based Topic Models

3.2.1 Label Aligned Neural Topic Model

In topic modeling, we often focus on the extent to which the discovered topics are meaningful and easy to understand, i.e., the interpretability of topics. Considering that the human-defined labels describe some inherent patterns within the documents, we expect to identify the relationships between labels and the discovered topics, namely, achieve **label-topic alignment**. Typical neural topic models like SCHOLAR use the latent document-topic distributions to predict the labels so as to achieve label-topic alignment indirectly. However, they are faced with the following challenges: (1) A label-free prior $p(z)$, which treats each topic evenly, is often adopted to generate the latent topics z_i for each word. As the generated topics are not specific to the label y, this will blend the semantics of multiple categories. (2) Under the label-free prior, they are not able to construct the label-topic relationships explicitly and thus fail to provide an intuitive view of the relationships between labels and topics. Given the challenges mentioned above, it is desirable that **a label-conditioned prior is utilized**. Motivated by this, we propose a novel supervised NTM, **Label Aligned Neural Topic Model (LANTM)**. Figure 3.12 shows the generative process of documents under (a) label-free prior and (b) label-conditioned prior, where w, z, y represent words, topics, and labels. Besides, different colors represent different labels.

3.2.1.1 Methodology

LANTM adopts a novel chain-structured graphical model that incorporates a label-conditioned prior. Furthermore, soft label-topic indicators are introduced and used in the prior/posterior. In order to obtain well-organized label-topic relationships, we formalize an entropy-regularized Optimal Transport (OT) problem on the embedding space and model the label-topic relationships as the transport plan. Besides, we propose to reconstruct the label-specific pseudo-documents to allow the label-topic distributions to capture the semantics of labels.

Chain-Structured Graphical Model We introduce the label guidance by setting the graphical model of LANTM to be a *chain*, i.e., $y \rightarrow z \rightarrow \boldsymbol{x}$. The generative process of LANTM can be described by $p(\boldsymbol{x}, y, z) = p(y)p(z|y)p(\boldsymbol{x}|y)$. We adopt a label-conditioned prior $p(z|y)$ to ensure that the latent topics will be specific to the label. This is different from most of previous works like SCHOLAR [8], where

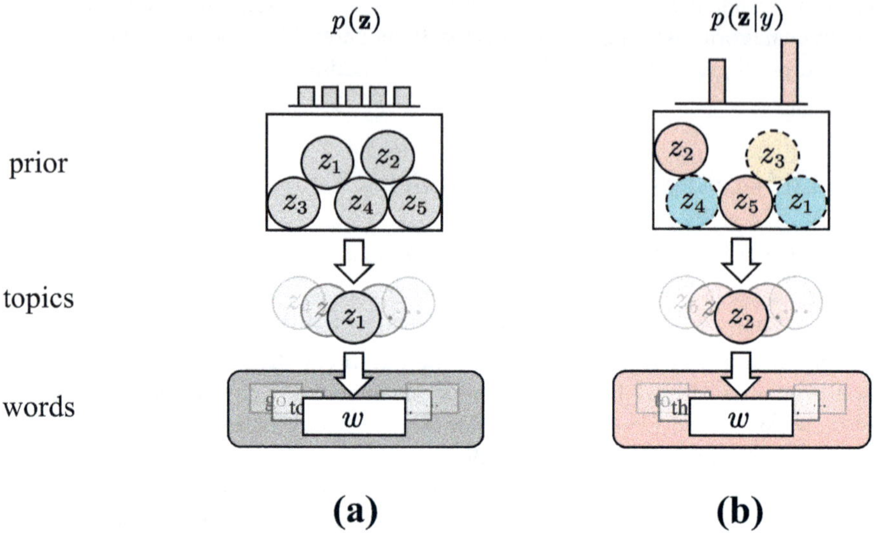

Fig. 3.12 The generative process of documents under (**a**) label-free prior $p(\mathbf{z})$ and (**b**) label-conditioned prior $p(\mathbf{z}|y)$

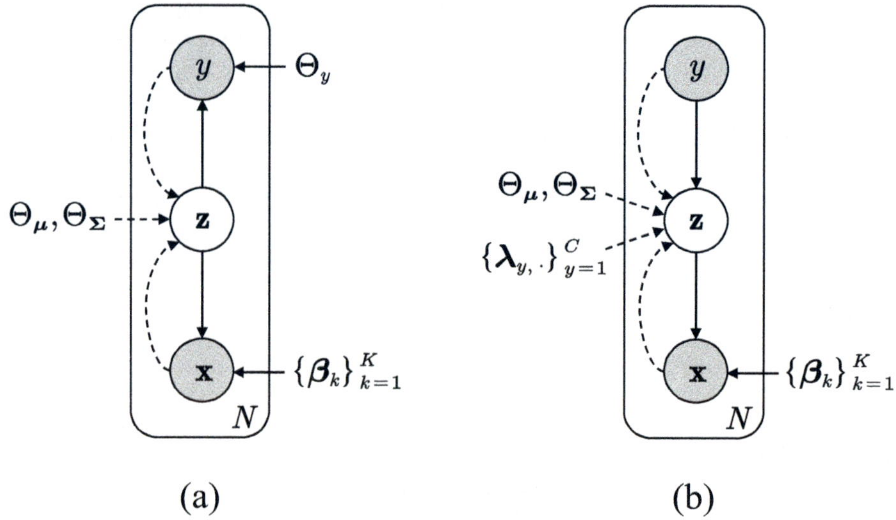

Fig. 3.13 Graphical models of (**a**) SCHOLAR and (**b**) LANTM (ours)

a label-free prior $p(z)$ is used. In this case, the posterior distributions of documents with different labels are driven toward the same label-free prior $p(z)$ (typically the isotropic normal distribution). For a certain topic, it will blend the semantics of multiple categories. Figure 3.13 presents the graphical models of SCHOLAR and

LANTM, where solid arrows denote generative process and dashed arrows denote inference. For our LANTM, we use the following objective function to maximize the ELBO:

$$\mathcal{L}(\boldsymbol{x}, y) = D_{\mathrm{KL}}[q(z|\boldsymbol{x}, y) \parallel p(z|y)] - \mathbb{E}_{q(z|\boldsymbol{x},y)}[\log p(\boldsymbol{x}|z)]. \tag{3.15}$$

Soft Indicators How to maximally distinguish documents with different labels? A natural idea is to ensure that each topic is activated only when a specific label is observed during inference. To this end, we introduce the label-topic soft indicators $\lambda_{y,k} \in (0, 1)$ to describe the relationship between label y and topic k. A value of $\lambda_{y,k} = 1$ means that topic k is fully activated, whereas $\lambda_{y,k} = 0$ indicates topic k is dummy under label y. With the help of soft indicators, we can model the label-conditioned prior $p(z|y)$ and posterior $q(z|\boldsymbol{x}, y)$ by

$$q(z|\boldsymbol{x}, y) = \mathcal{LN}(\boldsymbol{\mu} + \ln \boldsymbol{\lambda}_{y,\cdot}, \boldsymbol{\Sigma}),\ \ p(z|y) = \mathcal{LN}(\boldsymbol{\mu}_0 + \ln \boldsymbol{\lambda}_{y,\cdot}, \boldsymbol{\Sigma}_0), \tag{3.16}$$

where $\mathcal{LN}(\cdot, \cdot)$ denotes the logistic normal distribution. $\boldsymbol{\lambda}_{y,\cdot} = (\lambda_{y,1}, \ldots, \lambda_{y,K})$ is the indicator vector for y with respect to each topic.

Embedding-Based Label-Topic Relationship Modeling We denote the number of labels as C. For the label-topic indicator matrix $\boldsymbol{\lambda} = (\boldsymbol{\lambda}_{\cdot,1}, \ldots, \boldsymbol{\lambda}_{\cdot,K}) \in [0, 1]^{C\times K}$ where $\boldsymbol{\lambda}_{\cdot,k} = (\lambda_{1,k}, \ldots, \lambda_{C,k})$, we can set it as a hyperparameter or treat it as a learnable parameter. Considering the recent advance in topic modeling that uses the similarity between topic embeddings $\mathbf{T} = (\boldsymbol{t}_1, \ldots, \boldsymbol{t}_K)$ and word embeddings to compute the topic-word distributions $\{\boldsymbol{\phi}_k\}_{k=1}^K$, we also introduce label embeddings $\mathbf{L} = (\boldsymbol{l}_1, \ldots, \boldsymbol{l}_C)$ and construct label-topic relationships depending on $\mathbf{L}$ and $\mathbf{T}$.

For the learnable label-topic relationships, we expect the topics to capture the semantics specific to each label as much as possible. This means $\boldsymbol{\lambda}_{\cdot,k}$ is sparse enough. Besides, the emergence of super categories that dominate the majority of topics is undesirable, as it may do harm to the topic discovery of other categories. So we define $\boldsymbol{\lambda}$ as *well-organized* if:

1. (*Sparsity*) $\boldsymbol{\lambda}_{\cdot,k}^T \mathbf{1}_C = 1$ and $\boldsymbol{\lambda}_{\cdot,k}$ is sparse for all $k \in [1, 2, \ldots, K]$, where $\mathbf{1}_C$ is a C-dimensional all-one vector.
2. (*Balanced Assignment*) $N_y = \boldsymbol{\lambda}_{y,\cdot}^T \mathbf{1}_K$ is relatively balanced across different y's, where N_y is the "number" of topics assigned to label y.

Inspired by Transport Plan Dependency (TPD) proposed by Wu et al. [73], we formalize an entropy-regularized **Optimal Transport (OT) problem** [17] and model the label-topic relationships $\boldsymbol{\lambda}$ as the optimal transport plan $\mathbf{P}^*$ to obtain well-organized $\boldsymbol{\lambda}$. This is due to the following considerations: (1) $\mathbf{P}^*$ is sparse if the weight of entropic regularization term is small enough [17] and (2) $\mathbf{P}^*$ meets the preset marginal constraints which control the assignment.

Specifically, consider two probability measures: the source measure $\boldsymbol{\tau} = \sum_{y=1}^C a_y \delta_{\boldsymbol{l}_y}$ on label embeddings $\{\boldsymbol{l}_y\}_{y=1}^C$ and the target measure $\boldsymbol{\nu} = \sum_{k=1}^K b_k \delta_{\boldsymbol{t}_k}$ on topic embeddings $\{\boldsymbol{t}_k\}_{k=1}^K$, where δ_x denotes the Dirac delta function on x. Under a cost matrix $\mathbf{C} \in \mathbb{R}_{\geq 0}^{C\times K}$, we aim to find a transport plan $\mathbf{P}^* \in [0, 1]^{C\times K}$ that meets the marginal constraints and minimizes the objective function which includes both the total transport cost and an entropic regularization term:

$$\mathbf{P}^* = \underset{\mathbf{P}\in[0,1]^{C\times K}}{\arg\min} \; \langle \mathbf{P}, \mathbf{C}\rangle - \frac{1}{\epsilon}h(\mathbf{P})$$
$$\text{s.t.} \quad \mathbf{P}\mathbf{1}_K = \boldsymbol{a}, \mathbf{P}^T\mathbf{1}_C = \boldsymbol{b}, \tag{3.17}$$

where $\langle\cdot,\cdot\rangle$ denotes the Frobenius inner product. $1/\epsilon > 0$ is the weight of the entropic regularization term. $h(\mathbf{P})$ is the entropy of the joint distribution identified by $\mathbf{P}$. $\boldsymbol{a} = (a_1, \ldots, a_C)$ and $\boldsymbol{b} = (b_1, \ldots, b_K)$ are probability vectors corresponding to $\boldsymbol{\tau}$ and $\boldsymbol{\nu}$, respectively. By controlling $1/\epsilon$, we can obtain sparse $\mathbf{P}^*$. By letting $\boldsymbol{a}$ be the label distribution of the dataset and $\boldsymbol{b} = (1/K, \ldots, 1/K)$, we can ensure that the assignment is balanced.

It is natural to use the distance metric $\text{dist}(\cdot,\cdot)$ as the cost: $\mathbf{C}_{y,k} = \text{dist}(\boldsymbol{l}_y, \boldsymbol{t}_k)$. The choice of $\text{dist}(\cdot,\cdot)$ depends on how topic-word distributions are computed. For instance, ETM [20] utilizes inner product to measure the similarity between embeddings. In this situation, we define $\text{dist}(\boldsymbol{l}_y, \boldsymbol{t}_k) = \text{Softplus}(-\boldsymbol{l}_y^T\boldsymbol{t}_k)$. If the Gaussian kernel is used like ECRTM [70], then we may simply set $\text{dist}(\boldsymbol{l}_y, \boldsymbol{t}_k) = \|\boldsymbol{l}_y - \boldsymbol{t}_k\|_2^2$.

Pseudo-Documents Guided Label-Topic Relationship Learning We define normalized $\boldsymbol{\lambda}_{y,\cdot}$, namely, $\boldsymbol{\lambda}_{y,\cdot}/N_y$, as the *label-topic distribution* for y. We expect $\boldsymbol{\lambda}_{y,\cdot}$ to reliably describe the impact of each topic within label y. However, there is no connection between label-topic distributions and topic-word distributions, which causes the absence of the label semantics. To this end, we propose to reconstruct the y-specific pseudo-documents, denoted as $\boldsymbol{x}_{\text{pseu}}^y \in \mathbb{R}^V$, to enable the label-topic distribution to capture corresponding semantics and thus become more reasonable. Let $\mathcal{X}_y$ be the collection of documents with label y and $N_{\text{avg.}}$ be the average length of the documents across the whole corpus. We define $\boldsymbol{x}_{\text{pseu}}^y$ as follows:

$$\boldsymbol{x}_{\text{pseu}}^y = \frac{\sum_{\boldsymbol{x}\in\mathcal{X}_y}\boldsymbol{x}}{(\sum_{\boldsymbol{x}\in\mathcal{X}_y}\boldsymbol{x})^T\mathbf{1}_V} \cdot N_{\text{avg.}}. \tag{3.18}$$

Then, we use y's label-topic distribution $\boldsymbol{\lambda}_{y,\cdot}/N_y$ to reconstruct $\boldsymbol{x}_{\text{pseu}}^y$. We propose a novel loss function $\mathcal{L}_{\text{Sem.}}$, where $\Phi = (\boldsymbol{\phi}_1, \ldots, \boldsymbol{\phi}_K) \in \mathbb{R}^{V\times K}$:

$$\mathcal{L}_{\text{Sem.}} = -\mathbb{E}_y\left[(\boldsymbol{x}_{\text{pseu}}^y)^T \log\left(\Phi\frac{\boldsymbol{\lambda}_{y,\cdot}}{N_y}\right)\right]. \tag{3.19}$$

In practice, we find that the computation of the OT problem is difficult to converge when optimizing $\mathcal{L}(\boldsymbol{x}, y)$ and $\mathcal{L}_{\text{Sem.}}$, and the decrease of Sinkhorn distance [17] will alleviate it significantly. To this end, we minimize the Sinkhorn distance using $\mathcal{L}_{\text{OT}} = \langle\mathbf{P}^*, \mathbf{C}\rangle$. Finally, we adopt the following overall loss function:

$$\mathcal{L} = \frac{1}{|\mathcal{B}|} \sum_{(\boldsymbol{x},y)\in\mathcal{B}} \mathcal{L}(\mathbf{x}, y) + \gamma_1 \mathcal{L}_{\mathrm{OT}} + \gamma_2 \mathcal{L}_{\mathrm{Sem.}}, \tag{3.20}$$

where $\mathcal{B} \subset \mathcal{O}$ is the batch. γ_1 and γ_2 are the weights of $\mathcal{L}_{\mathrm{OT}}$ and $\mathcal{L}_{\mathrm{Sem.}}$, respectively.

Integrated with Other NTMs Note that our proposed model concentrates on inference rather than decoding. This implies that **our model can be combined with decoders of various existing NTMs**, such as ETM [20] and ECRTM [70]. For example, the decoder of ETM is parameterized by word and topic embeddings, and the topic-word distributions are computed by their inner products. We can attach ETM to LANTM by adopting the same way to compute the topic-word distributions.

3.2.1.2 Experimental Results

We conduct experiments on the following benchmark datasets: 20NewsGroup (**20NG**), **IMDB**, and Yahoo! Answers (**Yahoo**). We compare our LANTM with the following baseline methods: (1) **sLDA** [6], a classic supervised statistical topic model, (2) **SCHOLAR** [8], a classic supervised NTM, (3) **ETM** [20], and (4) **ECRTM** [70], two state-of-the-art unsupervised NTMs.

Label-Topic Alignment In this section, we examine the label-topic alignment achieved by our methods from the following aspects: (1) We visualize the label and topic embeddings and show the label-topic relationships for an intuitive view of label-topic alignment. (2) We conduct topic discovery on IMDB, where the human-defined labels do not describe the main patterns of the documents. (3) We perform clustering on the inferred document-topic distributions.

Visualization We run LANTM+ECRTM on 20NG to perform visualization. Figure 3.14a shows the visualization of the label embeddings and topic embeddings using t-SNE [48], as well as top-5 words of the topics under some labels. Topics under the same label are encircled together. And topics that are not explicitly assigned are circled in red, with their top-5 words presented in the red box in the upper right corner. Figure 3.14b is the heatmap of the label-topic indicator matrix, in which, the indices of topics without explicit assignment are highlighted in red. Figure 3.14 gives an intuitive view of label-topic alignment. From the visualization of the label and topic embeddings, we can see that modeling label-topic relationships on the embedding space leads to the refinement of the space, where topics under the same label are close to each other, while topics under different labels are far apart. Besides, we can observe from Fig. 3.14b that when modeling them as the transport plan, the label-topic relationships are well-organized.

Alignment Between Topics and Labels Describing Secondary Aspects The human-defined labels of the documents are sometimes not aligned with the co-occurrence patterns among words in the corpus. Take the sentiment labels in IMDB

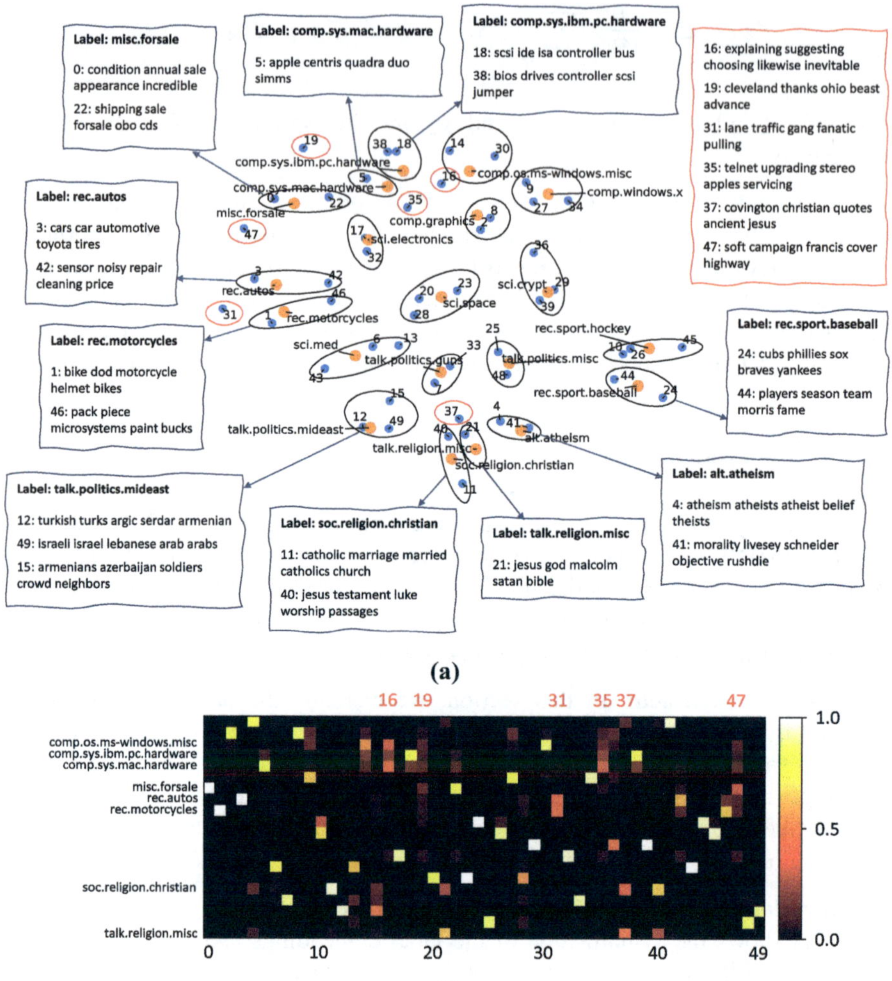

Fig. 3.14 Visualization of the discovered topics and the label-topic relationships on 20NG by LANTM+ECRTM with $K = 50$. (**a**) t-SNE visualization of label embeddings (*orange*) and topic embeddings (*blue*) (**b**) heatmap of the label-topic indicator matrix

dataset (positive and negative) for example. The sentiments of the audiences and descriptions of the film's content are two important patterns of the reviews in IMDB. It is unreasonable to assign any sentiment labels to the topics that fully describe the content of a film. Thanks to the label-topic indicators, LANTM is capable of discovering label-relevant and label-irrelevant topics simultaneously. Table 3.7 shows some most explicitly assigned and least explicitly assigned topics generated by LANTM+ECRTM on IMDB, where two sentiment labels, negative and positive, are defined. The value of polarity in this table is calculated by

Table 3.7 Some most explicitly assigned and least explicitly assigned topics discovered by LANTM+ECRTM on IMDB

Topic ID	Label	Polarity	Top-10 topic words
4	negative	0.489	waste horrible awful terrible crap worse sucked costs sucks crappy
12	positive	0.456	heartwarming tears april flawless sadness timeless compassion inspiring perfection unforgettable
26	positive	0.446	favorites fabulous bates nominated wonderful oscars flawless nomination wrenching delight
15	negative	0.367	waste flop travesty advise unlikeable choppy props checked offended awful
5	negative	0.329	hopelessly franco trashy hardcore walken boom preposterous tiresome overacting tame
37	negative	0.302	dialogue dialogues waste preview substance edgy distracting lacked poorly unlikeable
47	–	0.000	eddie murphy metal rock mario band school songs hello roll
17	–	0.000	games game saturday nicole russia friday minutes match bears night
8	–	0.006	martial chan jackie kung arts kong hong stunts jet chinese
31	–	0.012	films trailer waited spoof film powell park edward walked patience

$\max\{\lambda_{\text{positive},k}, \lambda_{\text{negative},k}\} - 0.5$ for the topic with ID k. The results show the meaningfulness of the label-topic indicators we propose in the scenario where the labels describe secondary aspects of the corpus.

Clustering-Based Evaluation We follow previous works [70] and calculate the following three cluster quality metrics: Adjusted Rand Index (ARI), Purity, and Normalized Mutual Information (NMI). We use two clustering strategies on the document-topic distributions of testing documents: (1) apply the **k-means** algorithm and (2) use the most dominant topic as the assigned cluster (referred to as the **top** strategy). The results are presented in Table 3.8. We run all models five times and report the averaged results. † indicates the gain of LANTM+ECRTM compared to supervised baseline methods is statistically significant at 0.05 level, while $*$ indicates the gain after applying LANTM is statistically significant at 0.05 level. The best scores are in bold. We do not report the results of sLDA on 20NG and Yahoo when $K = 100$ because it failed to converge in 48 hours. From the results we can see that our proposed methods can effectively enhance the label-topic alignment compared to both unsupervised NTMs and other supervised topic models.

Ablation Study In this section, we will discuss the impact of $\mathcal{L}_{\text{Sem.}}$. We control the value of γ_2 and run LANTM+ECRTM on 20NG by fixing $\gamma_1 = 5$. To have a more intuitive understanding about entropy of $\boldsymbol{\lambda}_{\cdot,k}$, Fig. 3.15 shows the heatmaps of the label-topic indicator matrix $\boldsymbol{\lambda}$ in four cases where γ_2 is set to 0, 0.1, 0.4, and 0.8, respectively. The *entropy* is the averaged value of $\boldsymbol{\lambda}_{\cdot,k}$, i.e.,

Table 3.8 ARI, Purity, and NMI scores under 50 and 100 topics

Dataset	Model	$K = 50$						$K = 100$					
		k-means			top			k-means			top		
		ARI	Purity	NMI	ARI	Purity	NMI	ARI	Purity	NMI	ARI	Purity	NMI
20NG	sLDA	†0.121	†0.457	†0.484	†0.324	†0.581	†0.490	–	–	–	–	–	–
	SCHOLAR	†0.040	†0.357	†0.391	†0.316	†0.629	†0.541	†0.011	†0.231	†0.276	†0.223	†0.637	†0.522
	ETM	*0.232	*0.492	*0.501	*0.265	*0.529	*0.429	*0.142	*0.420	*0.479	*0.186	*0.558	*0.430
	LANTM+ETM	0.333	0.584	0.560	0.312	0.624	0.507	**0.426**	**0.631**	**0.607**	0.223	0.684	0.521
	ECRTM	*0.101	*0.423	*0.466	*0.389	*0.574	*0.521	*0.049	*0.317	*0.377	*0.322	*0.559	*0.487
	LANTM+ECRTM	**0.370**	**0.595**	**0.578**	**0.415**	**0.635**	**0.573**	0.338	0.623	**0.607**	**0.444**	**0.689**	**0.599**
IMDB	sLDA	†0.033	†0.709	†0.082	†0.009	†0.676	†0.046	†0.026	†0.659	†0.063	†0.004	†0.658	†0.035
	SCHOLAR	†0.050	†0.675	†0.105	†0.017	**0.757**	0.086	†0.069	†0.666	†0.107	†0.010	**0.766**	**0.081**
	ETM	*0.020	*0.681	*0.057	*0.007	*0.656	*0.039	*0.039	*0.692	*0.076	*0.004	*0.664	*0.035
	LANTM+ETM	0.033	0.729	0.092	0.012	0.685	0.054	0.078	0.754	0.112	0.006	0.689	0.111
	ECRTM	*0.032	*0.667	*0.073	*0.013	*0.686	*0.055	*0.025	*0.656	*0.058	*0.005	*0.695	*0.046
	LANTM+ECRTM	**0.055**	**0.782**	**0.123**	**0.030**	0.742	**0.088**	**0.199**	**0.811**	**0.170**	**0.016**	0.762	0.080
Yahoo	sLDA	†0.084	†0.443	†0.312	†0.134	†0.517	†0.273	–	–	–	–	–	–
	SCHOLAR	†0.013	†0.287	†0.230	†0.113	†0.560	†0.317	†0.005	†0.199	†0.144	†0.070	†0.575	†0.313
	ETM	*0.117	*0.482	*0.310	*0.122	*0.502	*0.264	*0.068	*0.420	*0.303	*0.064	*0.496	*0.255
	LANTM+ETM	0.161	0.530	0.336	0.128	0.555	0.291	**0.222**	**0.549**	0.348	0.084	0.576	0.307
	ECRTM	*0.083	*0.455	*0.325	*0.130	*0.549	*0.299	*0.033	*0.319	*0.248	*0.103	*0.561	*0.304
	LANTM+ECRTM	**0.178**	**0.560**	**0.358**	**0.166**	**0.600**	**0.338**	0.177	0.541	**0.355**	**0.126**	**0.613**	**0.339**

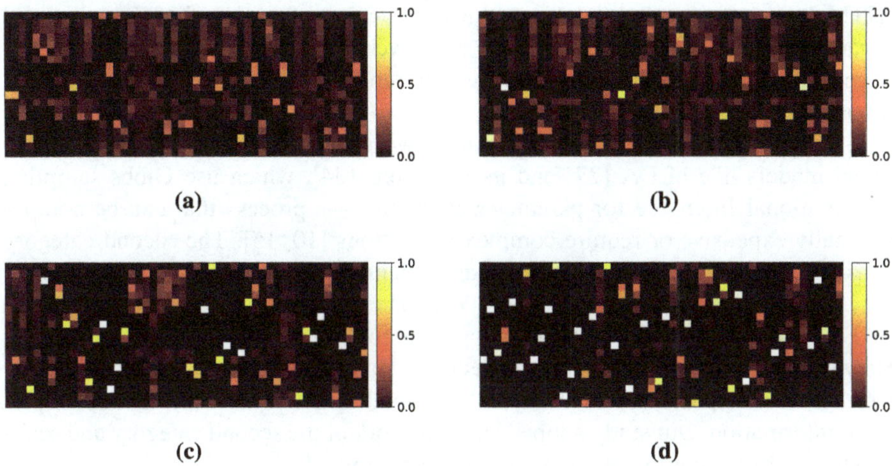

Fig. 3.15 Heatmaps of the label-topic indicator matrix $\boldsymbol{\lambda}$ learned by LANTM+ECRTM on 20NG with different γ_2 when $K = 50$. (**a**) $\gamma_2 = 0.0$, *entropy* $= 3.30$. (**b**) $\gamma_2 = 0.1$, *entropy* $= 3.01$. (**c**) $\gamma_2 = 0.4$, *entropy* $= 2.33$. (**d**) $\gamma_2 = 0.8$, *entropy* $= 1.64$

$entropy = \frac{1}{K}\sum_{k=1}^{K}\left(-\sum_{y=1}^{C}\lambda_{y,k}\log\lambda_{y,k}\right)$. We can observe from Fig. 3.15 that $\mathcal{L}_{\text{Sem.}}$ **has a huge impact on the learned structural characteristics of the label-topic relationships**. When $\mathcal{L}_{\text{Sem.}}$ is absent ($\gamma_2 = 0$), the label-topic relationships are fuzzy overall, as shown in Fig. 3.15a. Lacking guidance of the label semantics, the topics belong to many similar labels and lose specificity to labels. This is because under the objective of maximizing ELBO, the topics only capture patterns of word co-occurrence, which may be shared among similar labels. Meanwhile, $\mathcal{L}_{\text{Sem.}}$ is strong enough to amplify differences between labels. As γ_2 increases, the entropy of $\boldsymbol{\lambda}_{\cdot,k}$ decreases, which demonstrates well label-topic alignment.

3.2.2 Anchor Guided Hierarchical Topic Model

For the first time, we detect and add **anchor words** to the text generation process in an unsupervised way. And we enhance the model's representation of the anchor words through a **novel contrastive learning**.

3.2.2.1 Introduction

Topic models, capable of automatically uncovering coherent and meaningful topics from text collections, have been extensively applied in text analysis [33, 61, 69]. In these approaches, each topic is characterized by a set of relevant words representing a semantic concept. Unlike traditional flat topic models, Hierarchical Topic Models

(HTMs) aim to capture the hierarchical structure of topics for more rational organization [79]. HTMs have proven effective in tasks such as hierarchical web page classification [54] and discovering hierarchical relationships within academic repositories [56].

Existing HTMs can be categorized into two groups. The first includes conventional models like hLDA [27] and its variations [34], which use Gibbs sampling or Variational Inference for parameter estimation—a process that can be computationally expensive or require complex derivations [10, 15]. The second category consists of neural hierarchical topic models, including HNTM [14] and HyperMiner [75], among others [10, 15, 23, 31, 73]. These methods typically follow the VAE framework and utilize backpropagation for faster parameter inference [72]. However, most prior approaches neglect the significance of anchor words in the word generation process, potentially limiting their ability to mine fine-grained topic-word information. Our study adopts the framework of the second category and seeks to address this limitation by incorporating anchor words.

The concept of anchor word guidance [3] is grounded in the separability assumption [22], positing that each topic contains at least one highly relevant word that uniquely identifies it. For instance, while terms like "resurrection," "pray," "sin," and "christ" may relate to Christianity, only "christ" distinctly represents this topic. Based on this principle, there are two main types of anchor word-guided topic models: those that detect anchor words through an interactive process using nonnegative matrix factorization [2, 3, 53] and those that leverage label information [26, 32, 43]. While these methods highlight the importance of anchor words, they also face challenges related to computational stability and external dependency.

Despite their contributions, existing works on anchor words in topic modeling have limitations, particularly regarding unsupervised hierarchical topic mining. Two critical questions arise: First, how to detect anchor words without supervision in a neural hierarchical model, and second, how to enhance the representation of fine-grained topic-word relations and hierarchical structures using anchor words.

Our proposed **A**nchor Word Clustering and **P**ath Guided framework for unsupervised **H**ierarchical **T**opic **M**odeling (AP-HTM) addresses these issues by detecting anchor words via clustering algorithms [51, 74] and introducing four causal paths to guide the text generation process. Figures 3.16a, b illustrate the text generation processes of AP-HTM and other HTMs [10, 15, 40], where Θ is the set of parameters for the generation process, $T^{(l)}$ is the topic embedding at layer l, and z is the document-topic distribution. Besides N_d is the number of words in document d, W^A and W^* are the word embedding in anchor space and final space. Initially, we identify anchor words for each topic through clustering, forming a *topic →anchor word* path (Fig. 3.16c). Next, we incorporate two causal paths involving anchor words. To achieve a tightly integrated final embedding space, we introduce a causal path of *anchor word →word* (Fig. 3.16d), utilizing word co-occurrence graphs to capture relationships between anchor words and other words. Furthermore, we adopt a *topic+anchor word →higher layer topic* structure (Fig. 3.16e) to capture hierarchical topic relationships, leveraging anchor words as supplementary information. Finally, to decouple anchor words from other words in the causal

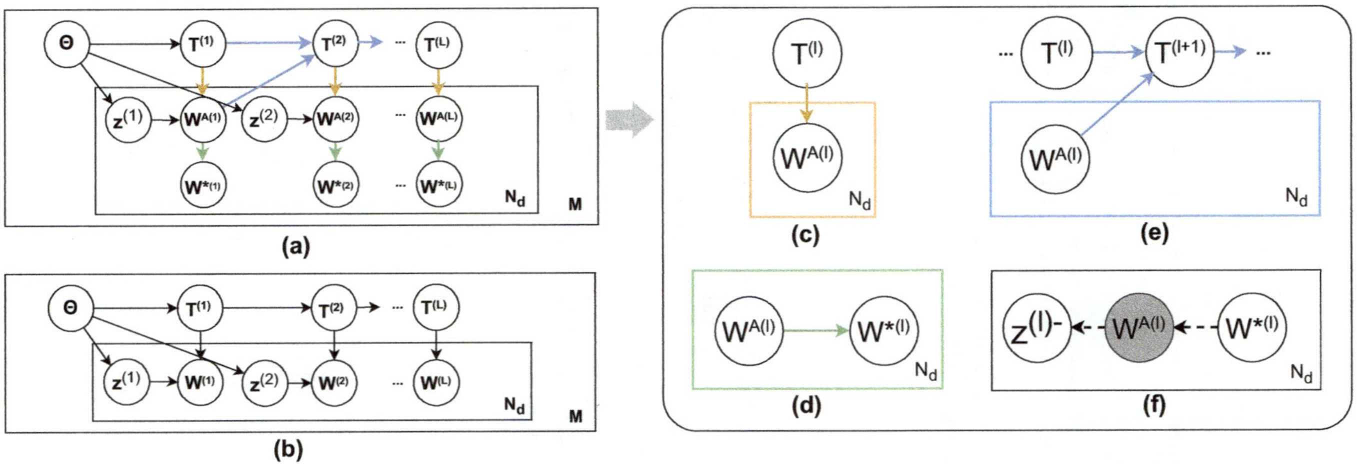

Fig. 3.16 The text generation process of (**a**) our AP-HTM and (**b**) other HTMs, (**c**) p_{clus}, (**d**) p_{intra}, and (**e**) p_{inter} show the decomposition of the three causal paths, and (**f**) p_{contra} demonstrates the effect of applying anchor perturbation (gray node) to W^A. The solid line represents the generation model, and the dotted line represents the inference process

diagram, we apply contrastive learning (Fig. 3.16f), masking anchor words during inference to focus the encoder on key aspects of anchor words.

In summary, our contributions are as follows:

- We propose a novel framework, AP-HTM (Fig. 3.17), which integrates anchor words as crucial elements into the text generation process. A clustering algorithm is adopted to adaptively detect these anchor words within the anchor space.
- Four causal path constraints are introduced into the VAE framework, guiding the identification of high-quality hierarchical topics.
- We introduce an innovative anchor-based contrastive learning approach that dynamically updates negative samples during the training process. This significantly enhances the model's representation capability for fine-grained anchor words.
- Extensive experiments were conducted on three datasets to evaluate our model. The results demonstrate that AP-HTM outperforms state-of-the-art baselines significantly.

3.2.2.2 Methodology

Anchor Space Clustering Following a previous work [51], we use the expectation-maximization (EM) algorithm to gradually sharpen the posterior topic-word distribution. In the E-Step, we estimate a new soft cluster assignment of each word based on the current parameters, and in the M-Step, we update the model parameters given the cluster assignments.

E-Step To estimate the cluster assignment of each word, we compute the posterior topic distribution of the kth topic at lth layer $t_k^{(l)}$ and ith word w_i obtained via the Bayes rule:

$$p\left(t_k^{(l)}|w_i\right) = \frac{p\left(w_i|t_k^{(l)}\right)p\left(t_k^{(l)}\right)}{\sum_{k'=1}^{K^{(l)}} p\left(w_i|t_{k'}^{(l)}\right)p\left(t_{k'}^{(l)}\right)}. \tag{3.21}$$

In the above equation, $p\left(w_i|t_k^{(l)}\right) = \phi_{k,i}^{(l)} = \exp\left(t_k^{(l)} \cdot w_i{}^T\right)$. And we assume that a topic $t_k^{(l)}$ is sampled from a uniform distribution over the $K^{(l)}$ topics, so $p\left(t_k^{(l)}\right) = \frac{1}{K^{(l)}}$. The posterior is simplified as $p\left(t_k^{(l)}|w_i\right) = \frac{\exp\left(t_k^{(l)} \cdot w_i{}^T\right)}{\sum_{k'=1}^{K} \exp\left(t_k^{(l)} \cdot w_i{}^T\right)}$.

Then we compute a new estimate of the cluster assignments $q(t_k|w_i^A)$ to be used for updating the model in the *M-Step* following a previous work [74]:

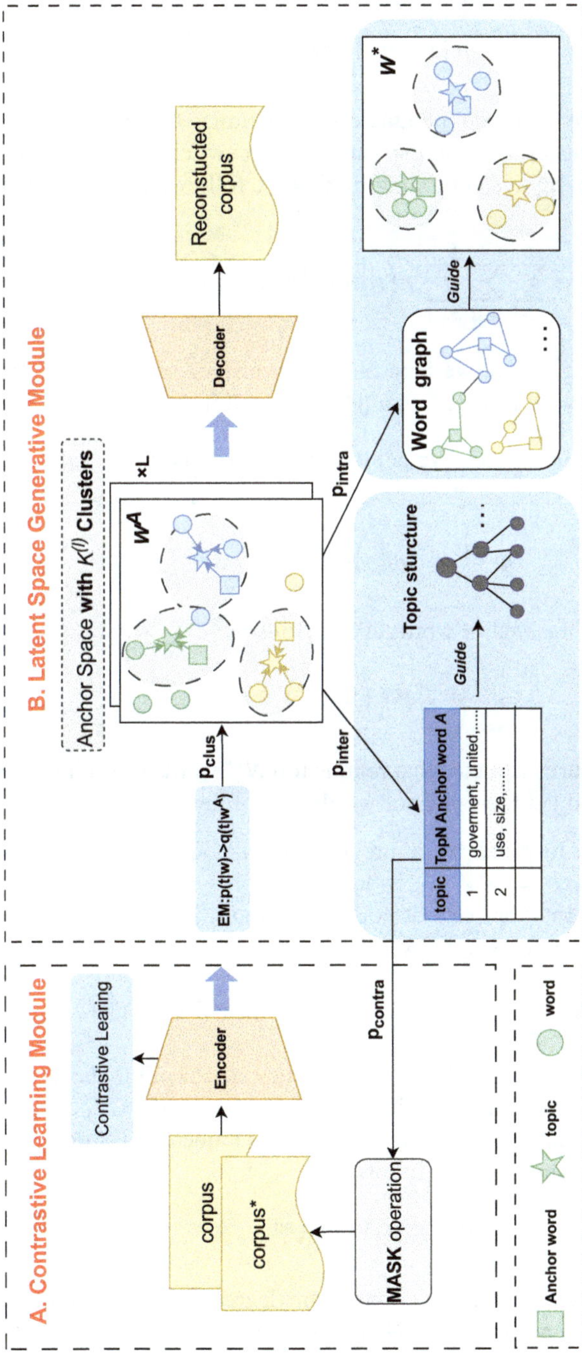

Fig. 3.17 The architecture of the proposed AP-HTM model

$$q\left(t_k^{(l)}|w_i{}^A\right) = \frac{p\left(t_k^{(l)}|w_i\right)^2/s_k^{(l)}}{\sum_{k'=1}^{K^{(l)}} p\left(t_{k'}^{(l)}|w_i\right)^2/s_{k'}^{(l)}}, \ s_k^{(l)} = \sum_{i=1}^{V} p\left(t_k^{(l)}|w_i\right). \tag{3.22}$$

M-Step We update the model parameters to maximize the expected log-probability of the current cluster assignment under the new cluster assignment estimate $\mathbb{E}_q[\log p]$, which is equivalent to minimizing the following cross-entropy loss:

$$\mathcal{L}_{\text{clus}} = -\sum_{l=1}^{L}\sum_{i=1}^{V}\sum_{k=1}^{K^{(l)}} q\left(t_k^{(l)}|w_i{}^A\right) \times \log p\left(t_k^{(l)}|w_i\right). \tag{3.23}$$

As shown in Fig. 3.16a, we here obtain the anchor space, where the words are embedded as $W^A = [w_1^A, w_2^A, \ldots, w_V^A]$ mentioned in p_{clus}.

Generating Anchor Words We get the topic-word distribution $\Phi^{A^{(l)}}$ at layer l in anchor space W^A by

$$\Phi^{A^{(l)}} = softmax\left(T^{(l)} \times (W^A)^T\right). \tag{3.24}$$

We then obtain the anchor words $A^{(l)}$ as follows:

$$A^{(l)} = TopN\left(\Phi^{A^{(l)}}\right) \in \mathbb{R}^{K^{(l)} \times N^w}, \tag{3.25}$$

where $TopN(\cdot)$ returns a vector that retains top N^w words of each row in Φ^A. Here we set N^w as 1 and get top-1 anchor words $A^{(l)}$ of layer l.

Intra-topic Path Inspired by a previous work [3], we construct a word co-occurrence graph $\mathcal{G} = (\mathcal{V}, \mathcal{E})$, where $\mathcal{V}$ (words as nodes) and $\mathcal{E}$ (counts of corresponding biterms) are sets of nodes and edges, respectively. We compute the modified adjacency matrix $\hat{\mathbf{A}}^{\mathbf{adj}}$ by

$$\hat{\mathbf{A}}^{\mathbf{adj}} = (\mathbf{D} + \mathbf{I}_N)^{-1/2}(\mathbf{A}^{adj} + \mathbf{I}_N)(\mathbf{D} + \mathbf{I}_N)^{-1/2}. \tag{3.26}$$

In the above, $\mathbf{A}^{\mathbf{adj}} \in \mathbb{R}^{V \times V}$ is the adjacency matrix, D is the degree matrix of $\mathbf{A}^{\mathbf{adj}}$, and I_N is the unit matrix. Unlike a previous work using GCNs [80] for learning the graph structure data, we achieve graph-guided embeddings through a simple and effective traction formulation as follows:

$$W^* = W^A + b\hat{\mathbf{A}}^{\mathbf{adj}} \times W^A, \tag{3.27}$$

where b is a hyperparameter that controls the degree of graph guidance. Intuitively, two word nodes that are similar in $\mathcal{G}$ are considered semantically related, which are therefore drawn closer to each other.

Inter-topic Path The fusion topic embedding $t_i^{*(l)}$ of ith topic at layer l can be achieved by

$$t_i^{*(l)} = \sigma\left(\mathbf{w}[t_i^{(l)}; A_i^{(l)}] + b\right), \tag{3.28}$$

where $\mathbf{w} \in \mathbb{R}^{2r \times r}$ and $b \in \mathbb{R}^r$ are the weights vector and bias, respectively, and $[\cdot;\cdot]$ represents the concatenation. The size of embedding vector remains unchanged after the fusion.

We use the EM algorithm to constrain hierarchical relationships between topics.

E-Step We have the relationship between ith topic of layer $l+1$ and jth topic of layer l as

$$p\left(t_j^{(l+1)} | t_i^{(l)}\right) = \Pi_{i,j}^{(l)}, i \in K^{(l)}, j \in K^{(l+1)}, \tag{3.29}$$

where $\Pi^{(l)} \in \mathbb{R}^{K^{(l)} \times K^{(l+1)}}$ is the matrix of learnable parameters that represents the topic hierarchy between layers l and $l+1$. Then we compute a posteriori estimation under p_{inter} as

$$q\left(t_j^{(l+1)} | t_i^{*(l)}\right) = \frac{\exp\left(t_i^{*(l)} \cdot t_j^{(l+1)^T}\right)}{\sum_{j'}^{K^{(l+1)}} \exp\left(t_i^{*(l)} \cdot t_{j'}^{(l+1)^T}\right)}. \tag{3.30}$$

M-Step We update the model parameters to encourage the topic hierarchy to utilize fine-grained anchor word information, which is equivalent to minimizing the following cross-entropy loss:

$$\mathcal{L}_{\text{inter}} = -\sum_{l=1}^{L-1} \sum_{j=1}^{K^{(l+1)}} \sum_{i=1}^{K^{(l)}} q\left(t_j^{(l+1)} \middle| t_i^{*(l)}\right) \times \log p\left(t_j^{(l+1)} \middle| t_i^{(l)}\right). \tag{3.31}$$

Negative Sampling We first obtain anchor words $A^{(l)}$ in Sect. 3.2.2.2 and then update the anchor word set by $S_A = \bigcup_{l \in L} A^{(l)}$. We believe that the anchor words are highly relevant to the semantics of the topic. Then, the mask operation is defined as removing words from S_A in the original document d to obtain a negative sample d^-. We enforce d^- to decouple the anchor word factor from d.

Positive Sampling Similar to the negative sampling strategy mentioned above, we obtain a set S_P of words with minimal relevance to all topics based on Φ^A. By removing the words in this set S_P from the original document d, we obtain d^+ as a positive sample of d. We believe that d^+ retains the salient topics from d. Let the document-topic distribution in layer l obtained from d, d^-, and d^+ be $z^{(l)}$, $z^{(l)^-}$, and $z^{(l)^+}$. We then calculate the contrastive loss as follows:

$$\mathcal{L}_{cl} = \sum_{l=1}^{L} \log \left(1 + \frac{\beta \cos(z^{(l)}, z^{(l)^-})}{\cos(z^{(l)}, z^{(l)+})} \right), \tag{3.32}$$

where β controls the weight of negative samples.

It is worth noting that under our generative framework, we adaptively update the negative and positive sampling during training, which is different from previous static contrastive learning.

3.2.2.3 Experimental Results

Interpretability of Topics The topic hierarchy generated by the model should exhibit two key properties: (1) a high degree of semantic consistency within individual topics and (2) a certain level of semantic similarity between parent and child topics. To evaluate these aspects, we utilize the widely adopted Normalized Pointwise Mutual Information (NPMI) [31] to assess the interpretability and coherence within topics and Cross-Level Normalized Pointwise Mutual Information (CLNPMI) [15] to measure the hierarchical relationship between parent and child topics.

As shown in Table 3.9, our proposed model significantly outperforms previous neural hierarchical topic models across all datasets, achieving improvements in NPMI by margins of 13.8%, 7.0%, and 15.8%, respectively, on three datasets. Moreover, while our model achieves the best CLNPMI score on the Wikitext-103 dataset, it also delivers competitive results on the other two datasets. Overall, for the CLNPMI metric, our model shows substantial improvement over the current optimal baseline models. This indicates that incorporating anchor words into the hierarchical generation process ensures coherence between parent and child topics, thus validating the structural rationality and effectiveness of the generated topic hierarchy.

Topic Diversity The diversity of hierarchical topics reflects the model's capability to uncover the richness and variety of information within a corpus. To evaluate the diversity of the generated hierarchical topics, we utilize Topic Uniqueness (TU) [55]. High TU scores indicate that the topics are distinct from each other, contributing to a more diverse set of topics.

As shown in Table 3.9, our proposed AP-HTM achieves the highest TU scores across all datasets. This superior performance can be attributed to two key components: p_{clus}, which clusters the anchor space W^A, allowing for a more refined topic representation, and p_{contra}, which enhances the encoder's ability to accurately capture the significance of anchor words. These mechanisms collectively ensure that the generated topics are both unique and meaningful, thus improving the overall diversity of the topic hierarchy.

Topic Quality Higher NPMI values typically signify stronger internal coherence within topics. However, this might also result in greater redundancy among topics, potentially lowering TU scores. Conversely, topics with elevated TU often represent

Table 3.9 Performance of all hierarchical topic models, where "–" means non-convergence after 48-hour training. The best results are in bold, and the second-best are underlined

Dataset	Metric	SawETM	HyperMiner	nTSNTM	nFNTM	CluHTM	TraCo	NG-HTM	AP-HTM
NIPS	NPMI ↑	0.133	0.135	0.100	0.113	0.137	0.112	0.147	**0.162**
	CLNPMI ↑	0.034	0.048	0.022	0.025	0.027	**0.054**	0.028	0.053
	TU ↑	0.431	0.662	0.373	0.765	0.554	0.672	0.719	**0.786**
	TQ ↑	0.057	0.089	0.037	0.086	0.076	0.076	0.106	**0.128**
20News	NPMI ↑	0.264	0.263	0.284	0.246	0.219	0.241	0.307	**0.329**
	CLNPMI ↑	0.138	0.153	0.156	0.150	0.164	0.129	0.146	**0.174**
	TU ↑	0.716	0.486	0.757	0.844	0.577	0.617	0.811	**0.857**
	TQ ↑	0.189	0.128	0.215	0.208	0.126	0.149	0.249	**0.281**
Wikitext-103	NPMI ↑	0.154	0.225	0.225	0.228	–	0.134	0.255	**0.295**
	CLNPMI ↑	0.060	0.079	0.121	**0.147**	–	0.042	0.090	0.145
	TU ↑	0.221	0.520	0.662	0.739	–	0.729	0.797	**0.825**
	TQ ↑	0.034	0.117	0.149	0.168	–	0.098	0.203	**0.243**

more peripheral subjects [71], which can lead to reduced NPMI scores. To address these considerations and offer a thorough assessment of topic quality, we utilize the Topic Quality (TQ) metric [20].

This metric aims to harmonize coherence within topics (assessed via NPMI) with distinctiveness across topics (evaluated through TU), thereby providing an integrated measure of overall topic excellence. As illustrated in Table 3.9, our proposed AP-HTM model surpasses all baseline models in terms of TQ scores. This outcome underscores that AP-HTM successfully detects both highly coherent and uniquely distinctive hierarchical topics.

Topic Structure Rationality In a well-structured hierarchy, topics at higher levels should exhibit broader semantics, whereas those closer to the bottom should be more specific. To quantify this characteristic, we use Topic Specialization (TS) [34], which is defined by the formula:

$$\mathrm{TS}(\Phi) = 1 - \cos(\Phi, \Phi_{\mathrm{Norm}}),$$

where Φ represents the topic-word distribution and Φ_{Norm} denotes the word distribution across the entire corpus. A higher TS score indicates greater specialization of the topic.

As illustrated in Fig. 3.18, AP-HTM demonstrates a coherent pattern of topic specialization across various datasets. Additionally, the relatively high TS values observed at each layer highlight our model's capability to identify more distinct topics at every level.

Examples of Anchor Word Guidance Our model integrates anchor words into the text generation process to capture fine-grained word information. To illustrate the impact of anchor words on p_{intra}, we present five examples of the top-5 topic words in Table 3.10. The first and second rows represent the anchor space W^A and the final embeddings W^*, respectively.

Additionally, we computed the relative increase in NPMI scores for each topic following the guidance by anchor words. These results highlight the effectiveness of incorporating anchor words in enhancing topic coherence.

From both quantitative and qualitative perspectives, anchor word guidance contributes to more coherent topics. Firstly, the NPMI score shows a significant improvement. Secondly, manual observation reveals that some words with lower relevance are substituted by more semantically consistent terms. For instance, in Topic:1_65, the words <u>*"Canada"*</u> and <u>*"Canadian"*</u> were found alongside terms related to *time*, which clearly do not align with the semantic context of these two words. This discrepancy may arise because p_{clus} captures the frequent co-occurrence of *"Canada"* with time-related terms.

However, with the guidance provided by p_{intra}, we leverage the synergy between anchor words and graph structures to ensure semantic consistency. Specifically, this approach helps replace less relevant words like *"Canada"* and *"Canadian"* with more appropriate terms, thereby enhancing the coherence and interpretability of the topic.

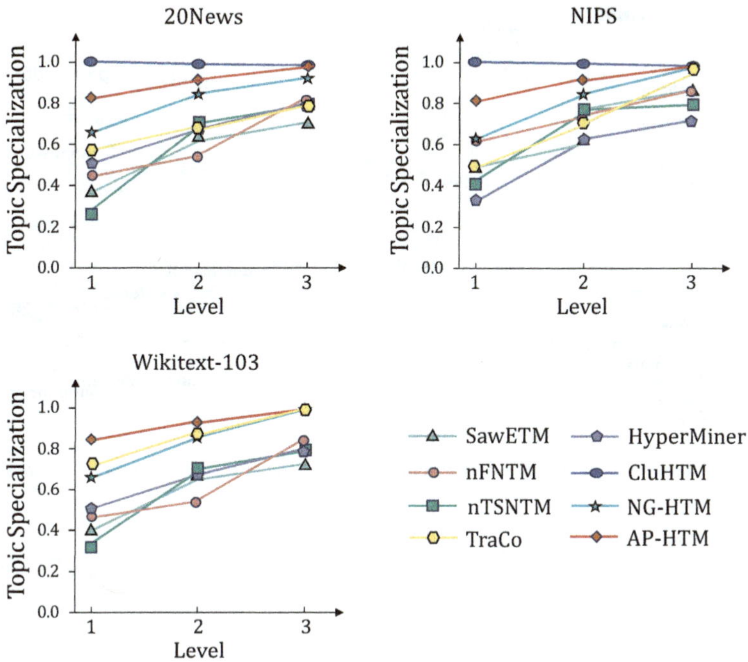

Fig. 3.18 Topic specialization of different topic structures across datasets

Table 3.10 Comparison of topic words before and after anchor word guidance. We italicize and underline words clearly unrelated to the topic

NPMI increase	Label	Top-5 words				
103%	topic:1_70	chastity	scripture	pray	sin	resurrection
		christ	jesus	bible	god	scriptures
86%	topic:1_65	yesterday	sunday	*Canada*	friday	*Canadian*
		sunday	saturday	thursday	yesterday	friday
59%	topic:1_14	interface	toolkit	unix	compiler	platforms
		interface	microsoft	unix	amiga	linux
50%	topic:1_111	father	son	woman	*pitt*	wife
		father	son	mother	woman	daughter
36%	topic:1_28	ordered	recommended	indicated	*initially*	plans
		ordered	announced	plans	announcement	recommended

Visualization of the Embedding Space To visually represent the top-5 words from five topics listed in Table 3.10, we utilize t-SNE visualization [48], as shown in Fig. 3.19. This visualization reveals that topics are embedded in the middle of related words, effectively conveying semantic information. Furthermore, words belonging to the same topic cluster more closely together, whereas those from different topics remain more dispersed.

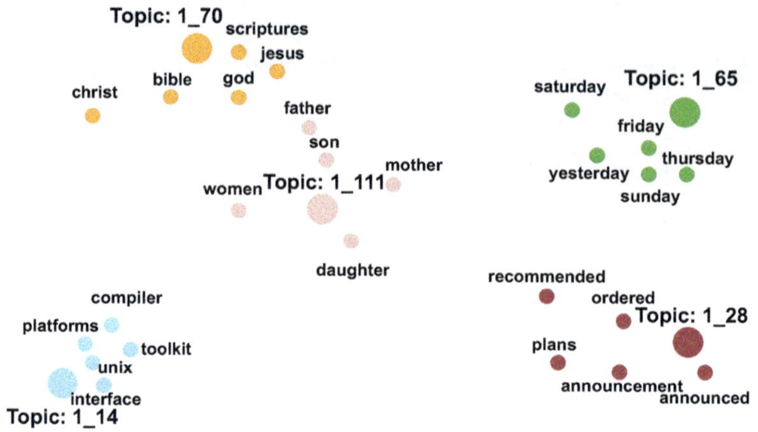

Fig. 3.19 Visualization of word and topic embeddings, where Topic: l_i denotes the ith topic at layer l

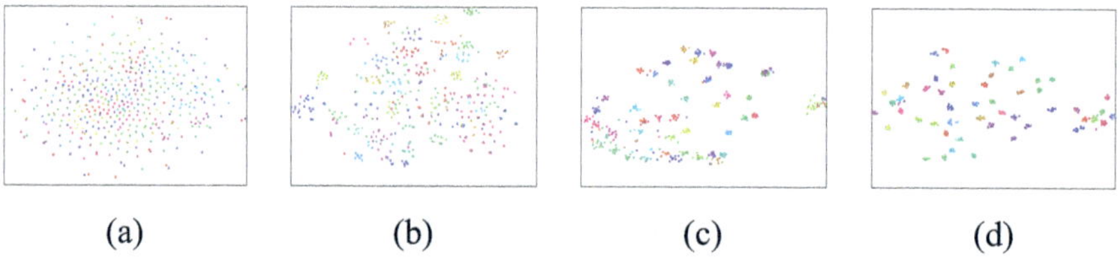

Fig. 3.20 Visualization of the embedding space for (**a**) TraCo, (**b**) HyperMiner, (**c**) NG-HTM, and (**d**) AP-HTM

Notably, semantically related topics are positioned closer to each other within the embedding space. For example, Topic:1_111 (*christianity*) and Topic:1_70 (*family*) exhibit proximity. Additionally, words with similar semantics across different topics tend to cluster near each other; for instance, *"god"*, *"jesus"* in Topic:1_111 and *"father"*, *"son"* in Topic:1_70 demonstrate this phenomenon.

The comparison between embedding spaces generated by AP-HTM and baseline models is illustrated in Fig. 3.20. We randomly select 50 topics from the third layer from 20News, each consisting of top-10 words assigned different colors. The figure highlights the superior clustering and semantic coherence achieved by our approach.

Visualization of the Topic Structure To intuitively demonstrate the capability of our model in generating coherent topic structures, we visualized several topic hierarchies extracted by AP-HTM from the 20News dataset. As illustrated in Fig. 3.21, each rectangle represents a topic along with its top-10 words, and arrows indicate the relationship from subtopics to their most relevant parent topics.

Consistent with the TS results presented in Fig. 3.18, topics exhibit a gradual semantic progression from general at the root to more specific toward the leaves. For example, *cancer* is a child topic under *health*, while *disk* is a child topic

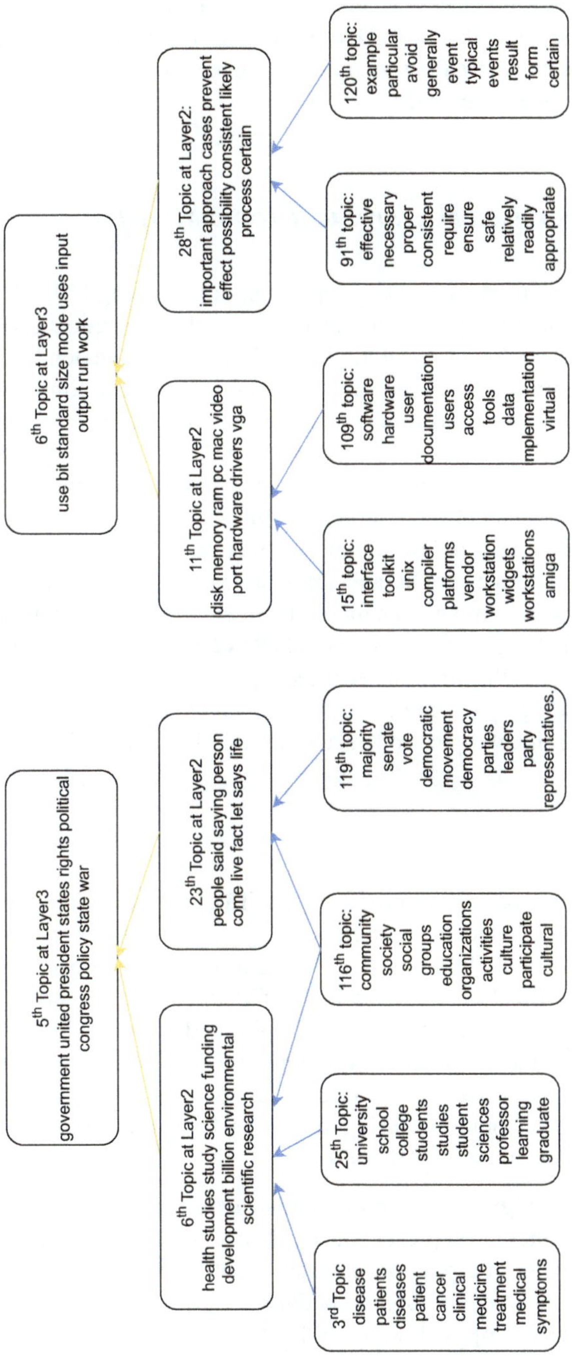

Fig. 3.21 An example of hierarchical topics learned from 20News by AP-HTM

under *use*. These relationships highlight that each topic's semantic meaning and the connections between adjacent layers are highly interpretable.

These findings suggest that AP-HTM effectively learns a reasonable and meaningful topic hierarchy, where the semantic coherence and relevance between parent and child topics are well-preserved.

3.2.3 Self-supervised Hierarchical Topic Model

With remarkable developments of text representation learning [19, 57, 65], mining topic hierarchy in the high-quality embedding space has become a promising idea. Particularly, the latest hierarchical topic models [15, 23] extend ETM [20] to hierarchical topic mining. However, most of these methods are limited in modeling semantic scopes of words and topics at different abstraction levels using classic point embeddings [57]. What is worst, they rely on the symmetric distance metric to infer the asymmetric hierarchy relations among topics. Such approximation results in an inaccurate hierarchical topic structure. Instead, geometric embeddings such as hyperbolic and box embeddings are more effective representations for structured data, including knowledge graphs and taxonomies [1, 5]. Considering the above challenges, we propose to mine topic hierarchy in the box embedding space [66] and develop a Box embedding-based topic model (**BoxTM**) following the framework of NVI. Figure 3.22 shows the differences of the topic hierarchy construction processes in the point embedding space and the box embedding space, which are adopted by most existing hierarchical topic models and our BoxTM, respectively.

3.2.3.1 Methodology

In this section, we introduce the proposed BoxTM in detail. Firstly, we introduce the background knowledge of box embeddings in the following section. Then we propose the box embedding-based document generative process, which is the main framework of BoxTM. In general, BoxTM infers topic distributions via the symmetric affinities and semantic scopes of topics and words in the box embedding space. Additionally, the hierarchical relations are modeled by the values of the asymmetric metric between topic boxes. Subsequently, we introduce the novel workflow of recursive topic clustering for upper-level topic mining. Finally, we introduce the learning strategy of BoxTM.

Background Knowledge As a representative geometric embedding technology, the box embedding method represents a word or topic as a box (i.e., axis-aligned hyperrectangle) instead of a point in the traditional Euclidean embedding method. With extra degrees of freedom, box embeddings can capture semantic scopes and asymmetric relations of objects [18, 39, 66].

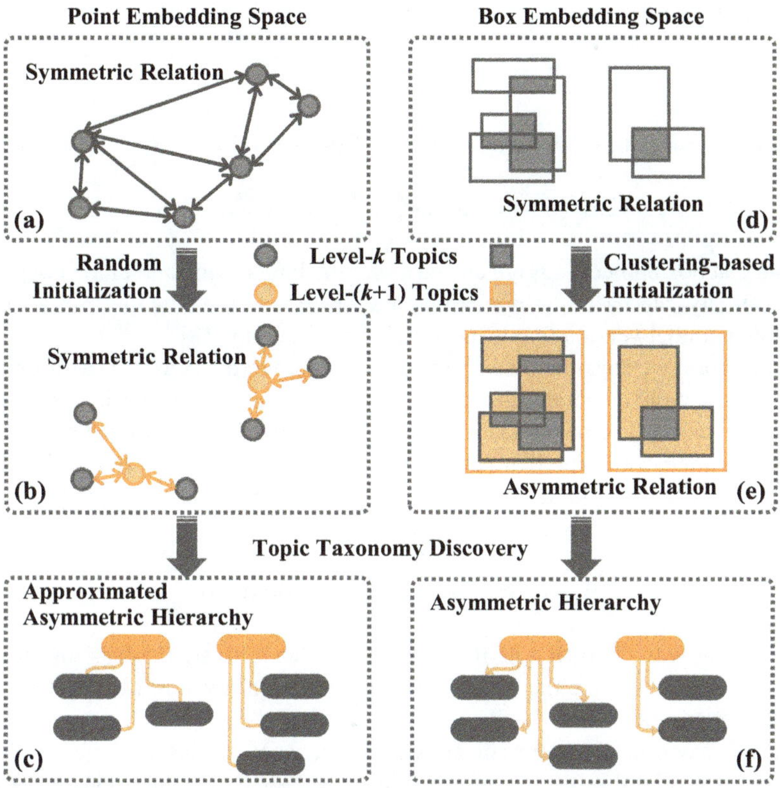

Fig. 3.22 Topic hierarchy construction in point (**a**–**c**) and box (**d**–**f**) embedding spaces for most existing hierarchical topic models and the proposed BoxTM

Definition 3.2 (Box Embedding) A D-dimensional box is determined by its minimum and maximum coordinates in each axis, parameterized by a pair of vectors $(\boldsymbol{x}_m, \boldsymbol{x}_M)$, where $\boldsymbol{x}_m, \boldsymbol{x}_M \in [0, 1]^D$ and $\boldsymbol{x}_{m,i} \leq \boldsymbol{x}_{M,i}$, for $\forall i \in \{1 \dots D\}$.

Definition 3.3 (Box Operations) Let $\text{Box}(A) := (\boldsymbol{x}_m^A, \boldsymbol{x}_M^A)$, $\text{Box}(B) := (\boldsymbol{x}_m^B, \boldsymbol{x}_M^B)$ denote box embeddings of objects A and B, respectively. The basic box operations are defined as follows:

Definition 3.3.1 (Volume) The volume of $\text{Box}(A)$ is defined as

$$\text{Vol}(\text{Box}(A)) := \prod_{i=1}^{D} (\boldsymbol{x}_{M,i}^A - \boldsymbol{x}_{m,i}^A). \tag{3.33}$$

Definition 3.3.2 (Intersection) If there is an overlap between $\text{Box}(A)$ and $\text{Box}(B)$, their intersection box is defined to be

$$\text{Box}(A) \wedge \text{Box}(B) := (\max(\boldsymbol{x}_m^A, \boldsymbol{x}_m^B), \min(\boldsymbol{x}_M^A, \boldsymbol{x}_M^B)); \tag{3.34}$$

otherwise, it is defined to be

$$\mathrm{Box}(A) \wedge \mathrm{Box}(B) := \perp . \tag{3.35}$$

Definition 3.3.3 (Union) The union box of $\mathrm{Box}(A)$ and $\mathrm{Box}(B)$ is defined as

$$\mathrm{Box}(A) \vee \mathrm{Box}(B) := (\min(\boldsymbol{x}_m^A, \boldsymbol{x}_m^B), \max(\boldsymbol{x}_M^A, \boldsymbol{x}_M^B)). \tag{3.36}$$

Note that box embeddings are closed under the intersection and union operations. For simplicity, the base box operations are described above, while in practice we adopt the Gumbel version that is more stable for training [18].

In this work, we consider the volume of a topic or word box as its size of semantic scope, i.e., a more general concept covers a larger region in the latent semantic space. The union box of topics and words is a generalization of their semantics. For the symmetric affinity, denoted as R_1, there is $\forall A, B : A R_1 B \Rightarrow B R_1 A$. We estimate R_1 with the volume of the intersection between topic and word boxes (R_s), which is defined as follows:

$$\mathrm{R}_s(A, B) = \mathrm{Vol}\left(\mathrm{Box}(A) \wedge \mathrm{Box}(B)\right). \tag{3.37}$$

Accordingly, we have $\mathrm{R}_s(A, B) = \mathrm{R}_s(B, A)$. To mitigate the bias toward large boxes, we can regularize the $\mathrm{R}_s(A, B)$ metric through division by $\mathrm{Vol}\left(\mathrm{Box}(A)\right) \cdot \mathrm{Vol}\left(\mathrm{Box}(B)\right)$ in practice.

For the asymmetric hierarchical relation between topics of adjacent levels, denoted as R_2, there is $\forall t^i, t^j \in \mathcal{T} : t^i R_2 t^j \Rightarrow \neg t^j R_2 t^i$, which means "if t^i is a sub-topic of t^j, then t^j is NOT a sub-topic of t^i." We reflect R_2 by the ratio of the volume of their intersection box to the upper-level topic box (R_a), that is,

$$\mathrm{R}_a\left(t_k^i \middle| t_{k+1}^j\right) = \frac{\mathrm{Vol}\left(\mathrm{Box}(t_k^i) \wedge \mathrm{Box}(t_{k+1}^j)\right)}{\mathrm{Vol}\left(\mathrm{Box}(t_{k+1}^j)\right)}, \tag{3.38}$$

where $t_k^i \in \mathcal{T}_k$ and $t_{k+1}^j \in \mathcal{T}_{k+1}$ denote topics of the kth and (k+1)th level, respectively. Unlike $\mathrm{R}_s(\cdot, \cdot)$, $\mathrm{R}_a(\cdot|\cdot)$ has the property that $\mathrm{R}_a(A|B) = \mathrm{R}_a(B|A) \neq 0$ iff $\mathrm{Vol}\left(\mathrm{Box}(A)\right) = \mathrm{Vol}\left(\mathrm{Box}(B)\right)$. Thus $\mathrm{R}_a(\cdot|\cdot)$ can better model the hierarchical relation that is asymmetric.

Document Generative Process BoxTM holds the assumption that a document is generated by any topics in the topic hierarchy and adopts a bottom-up hierarchical topic mining method. For NVI, BoxTM adopts a classic VAE with a logistic normal distribution $\mathcal{LN}(\mathbf{0}, \boldsymbol{I})$ [4] as the prior of topic proportion. A VAE consists of an encoder that learns hierarchical topic proportions given document representations and a decoder that reconstructs documents based on hierarchical topic proportions and topic distributions. The main framework of BoxTM is shown in Fig. 3.23.

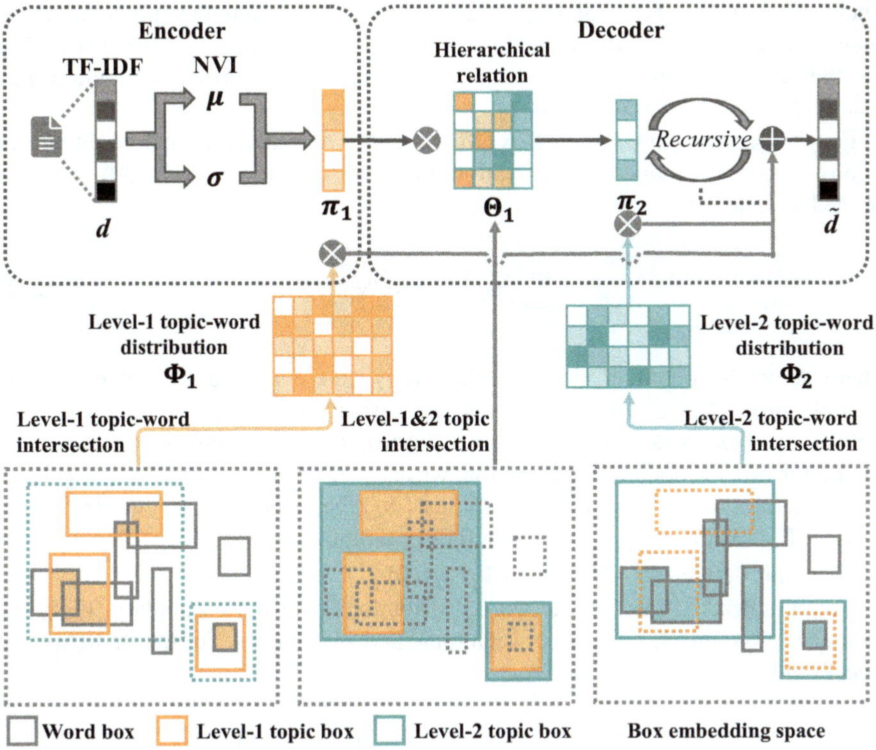

Fig. 3.23 The main framework of BoxTM

Given a corpus $\mathcal{D}$ and a vocabulary $\mathcal{V}$, BoxTM firstly encodes the TF-IDF representation $\boldsymbol{d} \in \mathbb{R}^{|\mathcal{V}|}$ of each document into a latent distribution, from which the latent feature z is sampled. After transforming z to acquire the leaf topic proportion $\boldsymbol{\pi}_1$, we infer upper-level topic proportions $\{\boldsymbol{\pi}_{>1}\}$ based on the asymmetric relations $\{\Theta_k\}$ of topics in the box embedding space. Specifically, $\Theta_k \in \mathbb{R}^{|\mathcal{T}_k|\times|\mathcal{T}_{k+1}|}$ between level-k topics $\mathcal{T}_k$ and the upper-level topics $\mathcal{T}_{k+1}$ are estimated by the asymmetric metric $\mathrm{R}_a(\cdot|\cdot)$, i.e.,

$$\Theta_k^{ij} = \log \mathrm{R}_a\left(t_k^i \middle| t_{k+1}^j\right), \tag{3.39}$$

where $t_k^i \in \mathcal{T}_k$ and $t_{k+1}^j \in \mathcal{T}_{k+1}$. The encoding process of BoxTM is defined as

$$\boldsymbol{h} = f_{\boldsymbol{h}}(\boldsymbol{d}), \tag{3.40}$$

$$z \sim \mathcal{N}\left(f_{\mu}(\boldsymbol{h}), f_{\sigma}(\boldsymbol{h})\right), \tag{3.41}$$

$$\boldsymbol{\pi}_1 = \mathrm{Softmax}\left(f_{\pi}(z)\right), \tag{3.42}$$

$$\boldsymbol{\pi}_{k+1} = \mathrm{Softmax}\left(\boldsymbol{\pi}_k \Theta_k\right), \tag{3.43}$$

where $f_h(\cdot)$, $f_\mu(\cdot)$, $f_\sigma(\cdot)$, and $f_\pi(\cdot)$ are MLPs. As the sampling process for the latent feature z is not differentiable, we adopt the reparameterization trick [59], and the sampled feature z can be expressed by a standard normal distribution, i.e., $z = f_\mu(\boldsymbol{h}) + \epsilon \cdot f_\sigma(\boldsymbol{h}), \epsilon \sim \mathcal{N}(\boldsymbol{0}, \boldsymbol{I})$.

For the decoding process of BoxTM, we apply normalization before document reconstruction to enhance the generation power of weak topic levels [30], which is defined as follows:

$$\tilde{\boldsymbol{d}} = \sum_{k=1}^{K} (\boldsymbol{\pi}_k \cdot \Phi_k) \circ \mathrm{CV}_{\Phi_k} / Z_k, \tag{3.44}$$

where K is the depth of the topic hierarchy and $\circ$ denotes the element-wise multiplication. $\Phi_k \in [0, 1]^{|\mathcal{T}_k| \times |\mathcal{V}|}$ is topic-word distributions of the kth level and $Z_k = ||(\boldsymbol{\pi}_k \cdot \Phi_k) \circ \mathrm{CV}_{\Phi_k}||_2$ is a 2-norm term. To weaken the impact of common words on document generation, we adopt the Coefficient of Variation (CV) [7] to sharpen all topic-word distributions $\{\Phi_k\}$. Specifically, the jth element of $\mathrm{CV}_{\Phi_k} \in \mathbb{R}^{|\mathcal{V}|}$ is the ratio of the standard deviation to the mean of the jth column in Φ_k, which is defined by $\mathrm{CV}^j_{\Phi_k} = \sigma(\Phi_k^{:,j})/\mu(\Phi_k^{:,j})$.

Notably, BoxTM infers topic-word distributions over the vocabulary $\mathcal{V}$ via the normalized symmetric affinity between topic and word boxes. For the ith topic t^i_k at level-k and the jth word w_j in $\mathcal{V}$,

$$\Phi_k^{ij} = \mathrm{Softmax}\left(\log \frac{\mathrm{R}_s(t^i_k, w_j)}{\mathrm{Vol}(t^i_k) \cdot \mathrm{Vol}(w_j)}\right), \tag{3.45}$$

which enables abstract topics to bias toward general words and vice versa.

Recursive Topic Clustering BoxTM conducts recursive clustering on topic boxes to learn upper-level topics to alleviate the problem of topic collapse, since the upper-level topic mining is guided by the correlation between lower-level topics. For the selection of clustering algorithms, we adopt the Affinity Propagation (AP) [25] algorithm for its flexibility and interpretability.

BoxTM constructs a topic affinity graph for topics at each level, where topic nodes are connected if their boxes overlap. However, the direct correlation between topics may be sparse in the box embedding space due to the diversity of topics, $\forall t^i_k, t^j_k \in \mathcal{T}_k$. To address this, we expand the semantic scope of each topic by merging the information of its keyword boxes. The box embedding of the processed ith topic $\tilde{t}^i_k$ at level-k is defined as follows:

$$\mathrm{Box}(\tilde{t}^i_k) := \left[\vee_{w \in \mathcal{W}^i_k} \mathrm{Box}(w)\right] \vee \mathrm{Box}(t^i_k), \tag{3.46}$$

where $\mathcal{W}^i_k = \{w_j | \arg\max_j \Phi_k^{ij}\}$ with $\left|\mathcal{W}^i_k\right| = n$ denotes the set of top-n ($n = 5$ in our experiments) representative words of topic t^i_k. Next, the affinity between topics is measured by the value of the asymmetric metric $\mathrm{R}_a(\cdot|\cdot)$ instead of the symmetric similarity metric $\mathrm{R}_s(\cdot,\cdot)$ because $\mathrm{R}_a(\cdot|\cdot)$ can weaken the influence of hub topics in clustering and prevent over-smoothing. Formally, the affinity matrix $\mathcal{A}_k \in \mathbb{R}^{|\mathcal{T}_k|\times|\mathcal{T}_k|}$ is defined by

$$\mathcal{A}_k^{ij} = \begin{cases} \log \mathrm{R}_a(\tilde{t}_k^j|\tilde{t}_k^i) & ,i \neq j; \\ 0 & ,i = j. \end{cases} \tag{3.47}$$

Later, the union of topic boxes in each cluster is adopted as a reasonable initialization of an upper-level topic. To reduce the impact of outliers in clustering, we propose a soft union operation $\vee_{\dagger}$, which is defined as follows:

$$\begin{aligned} \mathrm{Box}\left(t_{k+1}^i\right) &:= (x_m^i, x_M^i) = \vee_{\dagger t \in C_k^i} \mathrm{Box}(\tilde{t}), \\ x_m^i &= \mu(\{x_m^t\}_{t\in C_k^i}), x_M^i = \mu(\{x_M^t\}_{t\in C_k^i}), \end{aligned} \tag{3.48}$$

where C_k^i is the ith topic cluster of the kth level and $\mu(\cdot)$ is the mean operation. Besides, $\mathrm{Box}\left(t_{k+1}^i\right)$ is the reinitialized box embedding for the upper-level topic t_{k+1}^i. Then BoxTM infers the hierarchical relations Θ_k between level-k and level-(k+1) topics based on their box embeddings. For each topic $t_k^i \in \mathcal{T}_k$ at the kth level, its most relevant topic at the upper level is adopted as its parent topic $t_p^i \in \mathcal{T}_{k+1}$. Formally, we have

$$t_p^i := t_{k+1}^j = \arg\max_j \Theta_k^{ij}. \tag{3.49}$$

After conducting (K-1) times of topic clustering recursively, BoxTM can mine topics of K levels in a bottom-up manner.

Learning Strategy Similar to the training objective of VAEs, the main loss of BoxTM is to maximize the Evidence Lower BOund (ELBO). Specifically, the ELBO loss of BoxTM is defined as follows:

$$\mathcal{L}_{ELBO} = \mathbb{E}_{\pi_1\sim q_d}\log p(\boldsymbol{d}|\{\pi_k\},\{\Phi_k\}) - D_{KL}\left[q_d(\pi_1)||p(\pi_1)\right], \tag{3.50}$$

which balances between maximizing the expected log-likelihood (the first term) and minimizing the KL divergence (the second term) of the variational distribution $q_d(\pi_1) := \mathcal{N}(f_\mu(\boldsymbol{d}), f_\sigma(\boldsymbol{d}))$ and the prior distribution $p(\pi_1) := \mathcal{N}(\mathbf{0}, \boldsymbol{I})$.

For modeling the semantic scopes of words and topics, we propose two constraints, namely word-level constraint ($\mathcal{L}_{CO}$) and topic-level constraint ($\mathcal{L}_{HT}$). For the word-level constraint of semantic scope modeling, we adopt the cross-entropy loss $H(\cdot,\cdot)$ to constrain the relative volumes of word boxes among a randomly sampled batch $\mathcal{B} = \{(w_i, w_j)|P_{i|j} > 0\}$. Formally, we denote the box volume distribution as $q_{\mathrm{Box}}(w_i, w_j) \sim \mathrm{R}_a(w_i|w_j)$ and the co-occurrence pattern distribution as $p_{CO}(w_i, w_j) \sim P_{i|j}$. Then the loss function is defined by

$$\mathcal{L}_{CO} = H(p_{CO}, q_{\text{Box}}) = - \sum_{(w_i, w_j) \in \mathcal{B}} p_{CO}(w_i, w_j) \log q_{\text{Box}}(w_i, w_j). \tag{3.51}$$

For the topic-level constraint, we can define the following loss to maximize the score of asymmetric correlation metric between t_p and t_c:

$$\mathcal{L}_{HT} = - \sum_{(t_p, t_c) \in \mathcal{S}} \log \mathrm{R}_s(t_c, t_p) - \max\left[0, m - \log \mathrm{Vol}(t_p) + \log \mathrm{Vol}(t_c)\right]. \tag{3.52}$$

Accordingly, we define the regularization loss by

$$\mathcal{L}_{Box} = \alpha \cdot \mathcal{L}_{CO} + \beta \cdot \mathcal{L}_{HT}, \tag{3.53}$$

where α and β are weights for these losses. And the overall loss function of BoxTM is defined by

$$\mathcal{L} = \mathcal{L}_{ELBO} + \mathcal{L}_{Box}. \tag{3.54}$$

Then we adopt the Adam optimizer to update the network parameters of the encoder and box embeddings of topics and words. Based on the updated topic boxes, we perform a correction for the topic taxonomy using Eq. (3.49). Algorithm 10 shows the training workflow of the ith epoch in BoxTM. Intuitively, topic boxes

Algorithm 10: The ith epoch of training

Input: The corpus $\mathcal{D}$ and its vocabulary $\mathcal{V}$; The word and topic box embeddings W and $\{T_k\}$; The topic taxonomy $\mathcal{S}$ after prior epoch; The threshold γ for early stop.
Output: Updated word and topic box embeddings $\tilde{W}$ and $\{\tilde{T}_k\}$; Updated topic taxonomy $\tilde{\mathcal{S}}$.

if $i < \gamma$ **then**
  $\tilde{\mathcal{S}}, \{\tilde{T}_k\} \leftarrow$ RecurClus(W, T_1, K);
end
else
  $\tilde{\mathcal{S}}, \{\tilde{T}_k\} \leftarrow \mathcal{S}, \{T_k\}$;
end
for *each batch* $\mathcal{B} \subset \mathcal{D}$ **do**
  Infer hierarchical relations Θ by Eq. (3.39);
  Infer topic-word distributions Φ by Eq. (3.45);
  for *each document* $\boldsymbol{d} \in \mathcal{D}$ **do**
    Draw topic proportions $\{\boldsymbol{\pi}_k\} \leftarrow$ Encode($\boldsymbol{d}, \Theta$);
    Reconstruct document $\tilde{\boldsymbol{d}} \leftarrow$ Decode($\{\boldsymbol{\pi}_k\}, \Phi$);
  end
  Compute loss $\mathcal{L} = \mathcal{L}_{ELBO} + \mathcal{L}_{Box}$;
  Update $\tilde{W}$ and $\{\tilde{T}_k\}$ by minimizing $\mathcal{L}$;
  Update $\tilde{\mathcal{S}}$ based on $\{\tilde{T}_k\}$ by Eq. (3.49);
end
return $\tilde{W}, \{\tilde{T}_k\}, \tilde{\mathcal{S}}$

overlap less along with the training to capture diverse semantics, which limits the effectiveness of our recursive clustering module at the late phase of training. To tackle this problem, we use the early stopping trick that stops recursive clustering after the γth iteration. In our experiments, γ is set to 100.

3.2.3.2 Experimental Results

Datasets We conduct comprehensive evaluations on three benchmark datasets with latent topic hierarchies: (1) **20news**: A corpus consists of 20 newsgroups [62], (2) **NYT**: a set of news articles from the New York Times, which are categorized into 25 classes, and (3) **arXiv**: a set of paper abstracts covering 53 classes from arXiv website. The latter two datasets are collected by Meng et al. [50].

Baselines We compare our model with state-of-the-art topic taxonomy discovery models based on different frameworks, including document generation-based methods of **nTSNTM** [15], **SawETM** [23], **HyperMiner** [75], and **C-HNTM** [68], as well as a clustering-based method of **TaxoGen** [78]. Notably, HyperMiner adopts the hyperbolic embedding space, and the others hold the Euclidean embedding space assumption.

Intrinsic Evaluation of Topic Hierarchy For a reasonable topic hierarchy, each topic is a set of closely coherent words and diverse from one another. Besides, keywords of a parent topic t_p and its child topic t_c are coherent but have different semantic abstraction levels. Thus we validate the quality of the topic hierarchy from the following perspectives: (1) **Topic Coherence (C)**: We adopt a classic metric NPMI [35] to quantify the coherence of mined topics. (2) **Topic Diversity (D)**: The widely used TU [55] metric is for assessing the diversity among all topics, which is calculated by the number of unique keywords among all topics. (3) **Hierarchical Coherence (HC)**: We adopt the CLNPMI [15] metric to evaluate the hierarchical coherence between topics t_p and t_c.

Because highly overlapping topics may cause inflated coherence scores, the product of NPMI and TU is used as an integrated metric (**C*D**) for a comprehensive validation [20]. For the aforementioned metrics, we calculate the average of the scores of top-5, top-10, and top-15 topic words. Because the source code of nTSNTM and the algorithm of C-HNTM cannot adapt to topic taxonomy with more than three levels, their results on the arXiv dataset are not reported.

As shown in Table 3.11, BoxTM achieves new state-of-the-art results on most metrics across three datasets, when HyperMiner using hyperbolic embeddings outperforms SawETM. These results validate the advantage of geometric (i.e., hyperbolic and box) embeddings on hierarchical topic mining over traditional point embeddings. Compared to C-HNTM that performs poorly on the **HC** metric, the proposed recursive topic clustering module of BoxTM can effectively learn topics of different levels. While both SawETM and HyperMiner fail to learn a deep topic hierarchy on the arXiv dataset with massive documents, BoxTM remains outstanding performance on topic quality and hierarchical coherence. It validates

Table 3.11 Intrinsic metric scores on three datasets

Model	20news				NYT				arXiv			
	C	D	C*D	HC	C	D	C*D	HC	C	D	C*D	HC
nTSNTM	0.212	0.728	0.154	0.134	0.221	0.420	0.093	0.079	–	–	–	–
SawETM	0.221	0.404	0.089	0.098	0.228	0.476	0.109	0.084	0.134	0.256	0.034	0.047
HyperMiner	0.224	0.459	0.103	0.102	0.231	0.500	0.115	0.101	0.142	0.382	0.054	0.050
C-HNTM	0.196	0.633	0.124	0.090	0.152	0.458	0.070	0.036	–	–	–	–
TaxoGen	0.202	**0.789**	0.159	0.123	0.239	**0.881**	0.210	0.111	0.214	**0.681**	0.146	0.084
BoxTM	**0.301**	0.661	**0.199**	**0.159**	**0.409**	0.648	**0.265**	**0.177**	**0.257**	0.672	**0.173**	**0.113**

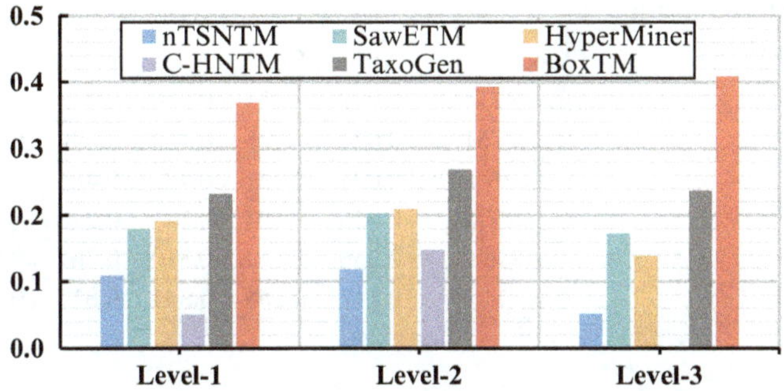

Fig. 3.24 The **C*D** scores at each level of BoxTM and baselines on NYT

that BoxTM not only has scalability for large-scale data but also has flexibility to learn topic hierarchies of different structures. In terms of the clustering-based method, TaxoGen obtains high scores of topic diversity (**D**) because each word only belongs to one topic at each level in its approach. However, it neglects the polysemy of some words, i.e., a word can be the keyword of different topics, which leads to its performance decline on topic coherence. For example, the word "*driver*" could be the keyword of topics "*hardware*" and "*motorcycles*."

Furthermore, Fig. 3.24 illustrates the **C*D** scores at each level of BoxTM and baselines on the NYT dataset. Both coherence and diversity of the level-2 topics of all models have different degrees of improvement compared to leaf topics. However, most baselines fail to learn high-quality topics at the root level, that is, they encounter the topic collapse problem. And topics mined by BoxTM remain high-quality at all levels, due to the effectiveness of the proposed recursive topic clustering module.

Extrinsic Evaluation of Topic Hierarchy As an important application scenario for hierarchical topic mining, the tree structure and keywords of the mined topic hierarchy can serve as auxiliary knowledge to improve the performance of hierarchical text clustering [36]. Specifically, each topic is regarded as a cluster, characterized by its keywords. We utilize the topic structure and the top-15 keywords of all topics learned by our BoxTM and baseline models as the inputs of a hierarchical text clustering model named WeSHClass [50]. For the evaluation metrics, we adopt two external criteria of clustering (i.e., **ARI** and $\mathbf{F}_\beta$) using golden labels (i.e., the manually annotated categories) of documents [63].

Table 3.12 shows the results of BoxTM and baseline models on the hierarchical text clustering task. Particularly, BoxTM and other HNTMs significantly outperform C-HNTM and TaxoGen that conduct clustering on word embeddings to mine topics, which reveals the limitation of latter methods in learning document-level semantics. Among HNTMs, BoxTM achieves the best results overall (**ARI** = 0.254 and $\mathbf{F}_\beta =$ 0.296 in average), followed by SawETM (**ARI** = 0.226 and $\mathbf{F}_\beta = 0.267$ in average). Although SawETM outperforms BoxTM on the arXiv dataset, it cannot discover coherent topics according to the intrinsic evaluation. These results show that there

Table 3.12 Extrinsic metric scores on three datasets

Model	20news		NYT		arXiv	
	ARI	$\mathbf{F}_\beta$	ARI	$\mathbf{F}_\beta$	ARI	$\mathbf{F}_\beta$
nTSNTM	0.081	0.133	0.389	0.448	–	–
SawETM	0.074	0.123	0.452	0.494	**0.151**	**0.184**
HyperMiner	0.075	0.127	0.421	0.466	0.115	0.151
C-HNTM	0.056	0.104	0.143	0.216	–	–
TaxoGen	0.066	0.132	0.310	0.367	0.097	0.133
BoxTM	**0.117**	**0.168**	**0.541**	**0.577**	0.103	0.143

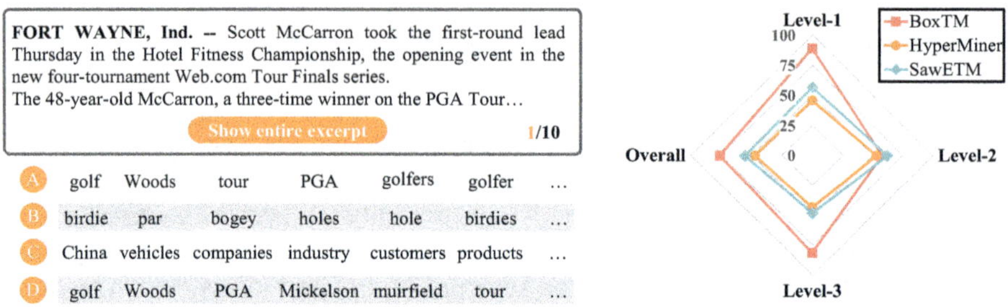

Fig. 3.25 Illustration of human evaluation on the NYT dataset: topic intrusion task example (left) and BoxTM and strong baselines' average precision (%) (right)

is a trade-off between learning high-quality topics and document-level semantics for topic modeling methods, and our BoxTM strikes a good balance.

Human Evaluation To complement the above automatic metrics, we also utilize a manual evaluation task of *topic intrusion* [9] to further validate the ability of topics at different levels to describe documents. As shown in Fig. 3.25 (left), human raters are shown a document from the testing set of NYT, along with four topics represented by their top-10 keywords. Three of them are the top-3 topics at the same level assigned to the given document by the topic model, while the remaining *intruder topic* is sampled randomly from the other low probability topics. We recruit ten graduate students majoring in computer science as raters and instruct them to choose topics that are not relevant to the documents. For evaluation, we compare our BoxTM with two strong baselines, i.e., SawETM and HyperMiner, excluding TaxoGen that cannot infer the topic distributions of documents. According to the value of Light's Kappa [41] ($\kappa = 0.607$), the annotation results of the ten raters have a fairly high degree of agreement.

Figure 3.25 (right) shows the precision scores of different models on this task. The performance of all three models on the manual assessment is generally consistent with those on the extrinsic evaluation. Notably, our BoxTM achieves an overall optimal result, which indicates that it generates different levels of topics that describe documents in alignment with human judgment.

References

1. Abboud R, Ceylan İİ, Lukasiewicz T, Salvatori T (2020) BoxE: a box embedding model for knowledge base completion. In: Advances in neural information processing systems, pp 9649–9661
2. Arora S, Ge R, Halpern Y, Mimno D, Moitra A, Sontag D, Wu Y, Zhu M (2013) A practical algorithm for topic modeling with provable guarantees. In: Proceedings of the 30th international conference on machine learning, pp 280–288
3. Arora S, Ge R, Moitra A (2012) Learning topic models–going beyond SVD. In: Proceedings of the 53rd annual IEEE symposium on foundations of computer science, pp 1–10
4. Atchison J, Shen SM (1980) Logistic-normal distributions: Some properties and uses. Biometrika 67(2):261–272
5. Bai Y, Ying Z, Ren H, Leskovec J (2021) Modeling heterogeneous hierarchies with relation-specific hyperbolic cones. In: Advances in neural information processing systems, pp 12316–12327
6. Blei DM, McAuliffe JD (2007) Supervised topic models. In: Advances in neural information processing systems, pp 121–128
7. Brown CE (1998) Coefficient of variation. In: Applied multivariate statistics in geohydrology and related sciences. Springer, Berlin, pp 155–157
8. Card D, Tan C, Smith NA (2018) Neural models for documents with metadata. In: Proceedings of the 56th annual meeting of the association for computational linguistics, pp 2031–2040
9. Chang JD, Boyd-Graber JL, Gerrish S, Wang C, Blei DM (2009) Reading tea leaves: how humans interpret topic models. In: Advances in neural information processing systems, pp 288–296
10. Chen H, Mao P, Lu Y, Rao Y (2023) Nonlinear structural equation model guided Gaussian mixture hierarchical topic modeling. In: Proceedings of the 61st annual meeting of the association for computational linguistics, pp 10377–10390
11. Chen R, Chen H, Lu Y, Rao Y, Zhu C (2025) Supervised neural topic modeling with label alignment. Trans Assoc Comput Linguist 13:249–263
12. Chen Y, Lei Z, Rao Y, Xie H, Wang FL, Yin J, Li Q (2022) Parallel non-negative matrix tri-factorization for text data co-clustering. IEEE Trans Knowl Data Eng 35(5):5132–5146
13. Chen Y, Wu J, Lin J, Liu R, Zhang H, Ye Z (2020) Affinity regularized non-negative matrix factorization for lifelong topic modeling. IEEE Trans Knowl Data Eng 32(7):1249–1262
14. Chen Z, Ding C, Rao Y, Xie H, Tao X, Cheng G, Wang FL (2021) Hierarchical neural topic modeling with manifold regularization. World Wide Web 24:2139–2160
15. Chen Z, Ding C, Zhang Z, Rao Y, Xie H (2021) Tree-structured topic modeling with nonparametric neural variational inference. In: Proceedings of the 59th annual meeting of the association for computational linguistics and the 11th international joint conference on natural language processing, pp 2343–2353
16. Chen Z, Liu B (2014) Topic modeling using topics from many domains, lifelong learning and big data. In: Proceedings of the 31th international conference on machine learning, pp 703–711
17. Cuturi M (2013) Sinkhorn distances: lightspeed computation of optimal transport. In: Advances in neural information processing systems, pp 2292–2300
18. Dasgupta SS, Boratko M, Zhang D, Vilnis L, Li X, McCallum A (2020) Improving local identifiability in probabilistic box embeddings. In: Advances in neural information processing systems, pp 182–192
19. Devlin J, Chang MW, Lee K, Toutanova K (2019) BERT: pre-training of deep bidirectional transformers for language understanding. In: Proceedings of the 2019 conference of the North American chapter of the association for computational linguistics: human language technologies, pp 4171–4186
20. Dieng AB, Ruiz FJ, Blei DM (2020) Topic modeling in embedding spaces. Trans Assoc Comput Linguist 8:439–453

21. Ding CHQ, Li T, Peng W, Park H (2006) Orthogonal nonnegative matrix t-factorizations for clustering. In: Proceedings of the 12th ACM SIGKDD international conference on knowledge discovery and data mining, pp 126–135
22. Donoho D, Stodden V (2003) When does non-negative matrix factorization give a correct decomposition into parts? In: Advances in neural information processing systems, pp 1141–1148
23. Duan Z, Wang D, Chen B, Wang C, Chen W, Li Y, Ren J, Zhou M (2021) Sawtooth factorial topic embeddings guided gamma belief network. In: Proceedings of the 38th international conference on machine learning, pp 2903–2913
24. Eshima S, Mochihashi D (2023) Scale-invariant infinite hierarchical topic model. In: Findings of the proceedings of the association for computational linguistics: ACL, pp 11731–11746
25. Frey BJ, Dueck D (2007) Clustering by passing messages between data points. Science 315(5814):972–976
26. Gallagher RJ, Reing K, Kale D, Ver Steeg G (2017) Anchored correlation explanation: topic modeling with minimal domain knowledge. Trans Assoc Comput Linguist 5:529–542
27. Griffiths T, Jordan M, Tenenbaum J, Blei D (2003) Hierarchical topic models and the nested Chinese restaurant process. In: Advances in neural information processing systems, pp 17–24
28. Gupta, P., Chaudhary, Y., Runkler, T., Schuetze, H.: Neural topic modeling with continual lifelong learning. In: Proceedings of the 37th International Conference on Machine Learning, pp 3907–3917 (2020)
29. Hu X, Wang R, Zhou D, Xiong Y (2020) Neural topic modeling with cycle-consistent adversarial training. In: Proceedings of the 2020 conference on empirical methods in natural language processing, pp 9018–9030
30. Hung WC, Jampani V, Liu S, Molchanov P, Yang MH, Kautz J (2019) SCOPS: Self-supervised co-part segmentation. In: Proceedings of the IEEE conference on computer vision and pattern recognition, pp 869–878
31. Isonuma M, Mori J, Bollegala D, Sakata I (2020) Tree-structured neural topic model. In: Proceedings of the 58th annual meeting of the association for computational linguistics, pp 800–806
32. Jagarlamudi J, Daumé III H, Udupa R (2012) Incorporating lexical priors into topic models. In: Proceedings of the 13th conference of the European chapter of the association for computational linguistics, pp 204–213 (2012)
33. Jelodar H, Wang Y, Orji R, Huang S (2020) Deep sentiment classification and topic discovery on novel coronavirus or covid-19 online discussions: NLP using LSTM recurrent neural network approach. IEEE J Biomed Health Inform 24(10):2733–2742
34. Kim JH, Kim D, Kim S, Oh A (2012) Modeling topic hierarchies with the recursive Chinese restaurant process. In: Proceedings of the 21st ACM international conference on information and knowledge management, pp 783–792
35. Lau JH, Newman D, Baldwin T (2014) Machine reading tea leaves: automatically evaluating topic coherence and topic model quality. In: Proceedings of the 14th conference of the European chapter of the association for computational linguistics, pp 530–539
36. Lee D, Shen J, Kang S, Yoon S, Han J, Yu H (2022) TaxoCom: topic taxonomy completion with hierarchical discovery of novel topic clusters. In: Proceedings of the ACM web conference, pp 2819–2829
37. Lee D, Shen J, Lee S, Yoon S, Yu H, Han J (2022) Topic taxonomy expansion via hierarchy-aware topic phrase generation. In: Findings of the proceedings of the association for computational linguistics: EMNLP
38. Lei Z, Liu H, Yan J, Rao Y, Li Q (2023) NMTF-LTM: towards an alignment of semantics for lifelong topic modeling. IEEE Trans Knowl Data Eng 35(10):10616–10632
39. Li X, Vilnis L, Zhang D, Boratko M, McCallum A (2019) Smoothing the geometry of probabilistic box embeddings. In: Proceedings of the 7th international conference on learning representations
40. Li Y, Wang C, Duan Z, Wang D, Chen B, An B, Zhou M (2022) Alleviating "posterior collapse" in deep topic models via policy gradient. In: Advances in neural information processing systems, pp 22562–22575

41. Light RJ (2011) Measures of response agreement for qualitative data: some generalizations and alternatives. Psychol Bull 76(5):365–377
42. Lim JP, Lauw H (2023) Large-scale correlation analysis of automated metrics for topic models. In: Proceedings of the 61st annual meeting of the association for computational linguistics, pp 13874–13898
43. Lin Y, Gao X, Chu X, Wang Y, Zhao J, Chen C (2023) Enhancing neural topic model with multi-level supervisions from seed words. In: Findings of the association for computational linguistics: ACL, pp 13361–13377
44. Liu J, Chen H, Zhu C, Rao Y (2024) Unsupervised hierarchical topic modeling via anchor word clustering and path guidance. In: Findings of the association for computational linguistics: EMNLP, pp 7505–7517
45. Liu Y, Zhang X, Zou M, Feng Z (2023) Co-occurrence embedding enhancement for long-tail problem in multi-interest recommendation. In: Proceedings of the 17th ACM conference on recommender systems, pp 820–825
46. Lu Y, Chen H, Mao P, Rao Y, Xie H, Wang FL, Li Q (2024) Self-supervised topic taxonomy discovery in the box embedding space. Trans Assoc Comput Linguist 12:1401–1416
47. Maas A, Daly RE, Pham PT, Huang D, Ng AY, Potts C (2011) Learning word vectors for sentiment analysis. In: Proceedings of the 49th annual meeting of the association for computational linguistics: human language Technology, pp 142–150
48. Van der Maaten L, Hinton G (2008) Visualizing data using t-SNE. J Mach Learn Res 9(11):2579–2605
49. Mehta SV, Patil D, Chandar S, Strubell E (2023) An empirical investigation of the role of pre-training in lifelong learning. J Mach Learn Res 24(214):1–50
50. Meng Y, Shen J, Zhang C, Han J (2019) Weakly-supervised hierarchical text classification. In: Proceedings of the 33rd AAAI conference on artificial intelligence, pp 6826–6833
51. Meng Y, Zhang Y, Huang J, Zhang Y, Han J (2022) Topic discovery via latent space clustering of pretrained language model representations. In: Proceedings of the ACM web conference, pp 3143–3152
52. Meng Y, Zhang Y, Huang J, Zhang Y, Zhang C, Han J (2020) Hierarchical topic mining via joint spherical tree and text embedding. In: Proceedings of the 26th ACM SIGKDD international conference on knowledge discovery & data mining, pp 1908–1917
53. Mimno D, Lee M (2014) Low-dimensional embeddings for interpretable anchor-based topic inference. In: Proceedings of the 2014 conference on empirical methods in natural language processing, pp 1319–1328
54. Ming ZY, Wang K, Chua TS (2010) Prototype hierarchy based clustering for the categorization and navigation of web collections. In: Proceedings of the 33rd international ACM SIGIR conference on research and development in information retrieval, pp 2–9
55. Nan F, Ding R, Nallapati R, Xiang B (2019) Topic modeling with Wasserstein autoencoders. In: Proceedings of the 57th annual meeting of the association for computational linguistics, pp 6345–6381
56. Paisley J, Wang C, Blei DM, Jordan MI (2014) Nested hierarchical Dirichlet processes. IEEE Trans Pattern Anal Mach Intell 37(2):256–270
57. Pennington J, Socher R, Manning CD (2014) Glove: global vectors for word representation. In: Proceedings of the 2014 conference on empirical methods in natural language processing, pp 1532–1543
58. Qin X, Lu Y, Chen Y, Rao Y (2021) Lifelong learning of topics and domain-specific word embeddings. In: Findings of the association for computational linguistics: ACL/IJCNLP, pp 2294–2309
59. Rezende DJ, Mohamed S, Wierstra D (2014) Stochastic backpropagation and approximate inference in deep generative models. In: Proceedings of the 31th international conference on machine learning, pp 1278–1286
60. Röder M, Both A, Hinneburg A (2015) Exploring the space of topic coherence measures. In: Proceedings of the 8th ACM international conference on web search and data mining, pp 399–408

61. Rubin TN, Chambers A, Smyth P, Steyvers M (2012) Statistical topic models for multi-label document classification. Mach Learn 88:157–208
62. Song Y, Roth D (2014) On dataless hierarchical text classification. In: Proceedings of the 28th AAAI conference on artificial intelligence, pp 1579–1585
63. Steinbach M, Kumar V, Tan P (2005) Cluster analysis: basic concepts and algorithms. In: Introduction to data mining, 1st edn. Pearson Addison Wesley
64. Vidhya M, Aji S (2022) Parallelized extreme learning machine for online data classification. Appl Intell 52(12):14164–14177
65. Vilnis L (2021) Geometric representation learning. Doctoral Dissertation, University of Massachusetts Amherst
66. Vilnis L, Li X, Murty S, McCallum A (2018) Probabilistic embedding of knowledge graphs with box lattice measures. In: Proceedings of the 56th annual meeting of the association for computational linguistics, pp 263–272
67. Wang C, Liu X, Song Y, Han J (2015) Towards interactive construction of topical hierarchy: a recursive tensor decomposition approach. In: Proceedings of the 21th ACM SIGKDD international conference on knowledge discovery and data mining, pp 1225–1234
68. Wang N, Wang D, Jiang T, Du C, Fang C, Zhuang F (2023) Hierarchical neural topic model with embedding cluster and neural variational inference. In: Proceedings of the 2023 SIAM international conference on data mining, pp 936–944
69. Wang W, Gan Z, Wang W, Shen D, Huang J, Ping W, Satheesh S, Carin L (2018) Topic compositional neural language model. In: Proceedings of the international conference on artificial intelligence and statistics, pp 356–365
70. Wu X, Dong X, Nguyen TT, Luu AT (2023) Effective neural topic modeling with embedding clustering regularization. In: Proceedings of the 40th international conference on machine learning, pp 37335–37357
71. Wu X, Li C, Zhu Y, Miao Y (2020) Short text topic modeling with topic distribution quantization and negative sampling decoder. In: Proceedings of the 2020 conference on empirical methods in natural language processing, pp 1772–1782
72. Wu X, Nguyen T, Luu AT (2024) A survey on neural topic models: methods, applications, and challenges. Artif Intell Rev 57(2):1–30
73. Wu X, Pan F, Nguyen T, Feng Y, Liu C, Nguyen CD, Luu AT (2024) On the affinity, rationality, and diversity of hierarchical topic modeling. In: Proceedings of the 38th AAAI conference on artificial intelligence, pp 19261–19269
74. Xie J, Girshick R, Farhadi A (2016) Unsupervised deep embedding for clustering analysis. In: Proceedings of the 33nd international conference on machine learning, pp 478–487
75. Xu Y, Wang D, Chen B, Lu R, Duan Z, Zhou M (2022) HyperMiner: topic taxonomy mining with hyperbolic embedding. In: Advances in neural information processing systems, pp 31557–31570
76. Yan J, Lu Y, Chen H, Yu J, Rao Y (2024) Lifelong hierarchical topic modeling via nonparametric word embedding clustering. In: Proceedings of the European conference on machine learning and knowledge discovery in databases, pp 270–287
77. Zhai M, Chen L, Tung F, He J, Nawhal M, Mori G (2019) Lifelong GAN: continual learning for conditional image generation. In: Proceedings of the 2019 IEEE/CVF international conference on computer vision, pp 2759–2768
78. Zhang C, Tao F, Chen X, Shen J, Jiang M, Sadler B, Vanni M, Han J (2018) TaxoGen: unsupervised topic taxonomy construction by adaptive term embedding and clustering. In: Proceedings of the 24th ACM SIGKDD international conference on knowledge discovery & data mining, pp 2701–2709
79. Zhang Z, Zhang X, Rao Y (2022) Nonparametric forest-structured neural topic modeling. In: Proceedings of the 29th international conference on computational linguistics, pp 2585–2597
80. Zhu Q, Feng Z, Li X (2018) GraphBTM: graph enhanced autoencoded variational inference for biterm topic model. In: Proceedings of the 2018 conference on empirical methods in natural language processing, pp 4663–4672

Chapter 4
Applications

Abstract Topic discovery and emotion detection are closely related such that public emotions always vary from one topic to another, and topics trigger public emotions. For instance, topics with representative words such as "care," "death," and "discover" are unvaried clues that, respectively, evoke emotions of "touching," "sadness," and "surprise" in users. However, there are also topics that may trigger different emotions among user groups as characterized by stance, gender, region, education, and others. This phenomenon is quite common in examples such as after a football match ends, the match result will evoke different emotions among fans from the two opposing sides. In light of this consideration, we propose a method of topic discovery and emotion detection with the help of user characteristics. To reduce noise in abundant features, we further propose two fast supervised topic models that are equipped with an accelerated algorithm. Finally, we develop an aspect extraction approach guided by document-topic distributions derived from neural topic models. This chapter will describe our work on topic-level emotion detection and topic-aware aspect extraction. For the former, we introduce topic-level emotion models exploiting user groups and labels, respectively. For the latter, we present a topic-aware dynamic convolutional neural network for aspect extraction.

4.1 Topic-Level Emotion Detection

In this section, we will introduce our work entitled "User Group Based Emotion Detection and Topic Discovery over Short Text" [2] and "Fast Supervised Topic Models for Short Text Emotion Detection" [7], which were developed for topic-level emotion detection by modeling topics and emotions jointly.

4.1.1 User Group-Based Topic-Emotion Model

Traditional topic-emotion modeling approaches often suffer from feature sparsity when dealing with short texts and fail to effectively capture user characteristics

Y. Rao, Q. Li, *Topic Modeling*, Machine Learning: Foundations, Methodologies, and Applications, https://doi.org/10.1007/978-981-96-8853-1_4

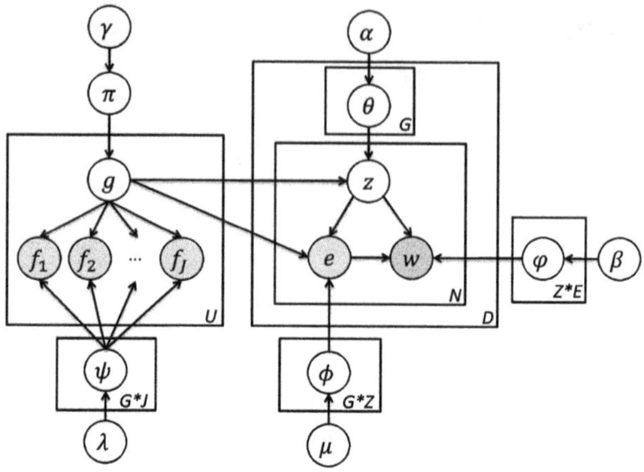

Fig. 4.1 Graphical model of UGTE. Shaded circles are observed variables and unshaded are latent. Plates show replication

in the emotion detection process. This limitation prevents them from discovering coherent group-level patterns and user-specific emotional tendencies.

To address this issue, we propose the User Group-based Topic-Emotion (UGTE) model, which jointly models topics, emotions, and user characteristics at the group level. Given a set of documents $D = \{d_1, d_2, \ldots, d_{|D|}\}$ with vocabulary $W = \{w_1, w_2, \ldots, w_{|W|}\}$, emotion labels $E = \{e_1, e_2, \ldots, e_{|E|}\}$, and users $U = \{u_1, u_2, \ldots, u_{|U|}\}$, each document d_i is generated by user u_r and labeled with emotion e_k, denoted as $d_i^{r,k}$. For user characteristics, we denote the set of characteristics tags of the jth type as $F_j = \{f_{j,1}, f_{j,2}, \ldots, f_{j,|F_j|}\}$.

The graphical representation of our model is shown in Fig. 4.1, which illustrates the hierarchical dependencies between user characteristics, groups, topics, and emotions. The left plate represents the iteration over different users, while the right plate shows the generation process for each word in the document.

Our model introduces a user group layer where each group g is associated with its own characteristics tag distributions $\psi_{g,j}$, topic distribution θ_g, and emotion distribution $\phi_{g,z}$. The key probability distributions in the generative process are defined as follows:

$$\pi_g = \frac{N_g + \gamma}{\sum_{g'=1}^{|G|}(N_{g'} + \gamma)}, \tag{4.1}$$

$$\psi_{g,j,f} = \frac{N_{g,f,j} + \lambda}{\sum_{f'=1}^{|F_j|}(N_{g,f',j} + \lambda)}, \tag{4.2}$$

$$\theta_{g,z} = \frac{N_{g,z} + \alpha}{\sum_{z'=1}^{|Z|}(N_{g,z'} + \alpha)}, \tag{4.3}$$

$$\phi_{g,z,e} = \frac{N_{g,z,e} + \mu}{\sum_{e'=1}^{|E|}(N_{g,z,e'} + \mu)}, \tag{4.4}$$

where N_g is the number of users assigned to group g, $N_{g,f,j}$ is the number of characteristics tags f_j assigned to group g, $N_{g,z}$ is the number of words assigned to topic z in group g, and $N_{g,z,e}$ is the number of words assigned to emotion e of topic z in group g. The hyperparameters involve γ, λ, α, and μ.

For parameter inference, we employ Gibbs sampling due to its simplicity. The posterior probability for inferring the group g_r of a user u_r can be derived by marginalizing the joint probability:

$$p(g_r = g|F^{u_r}, d_i^{r,k}, g_{-r}, \alpha, \beta, \mu, \lambda, \gamma) \propto$$

$$\frac{N_{-r}^{g} + \gamma}{\sum_{g'=1}^{|G|}(N_{-r}^{g'} + \gamma)} \times \prod_{j=1}^{J} \frac{N_{f,j}^{g,-r} + \lambda}{\sum_{f'=1}^{|F_j|} N_{f',j}^{g,-r} + |F_j|\lambda} \times$$

$$\prod_{n=1}^{N_i} \sum_{z=1}^{|z|} \frac{N_z^{-i,g} + \alpha}{\sum_{z'=1}^{|Z|}(N_{z'}^{-i,g} + \alpha)} \times \frac{N_{e_k}^{-i,g,z} + \mu}{\sum_{e'=1}^{|E|}(N_{e'}^{-i,g,z} + \mu)} \times \frac{N_{w_{i,n}}^{-i,z,e} + \beta}{\sum_{w'=1}^{|W|}(N_{w'}^{-i,z,e} + \beta)}. \tag{4.5}$$

For emotion prediction of unlabeled documents, we compute

$$p(e_{test}|d_{test}) = p(e_{test} = e|d_{test}, F^u) = \prod_{n=1}^{N_{test}} \sum_{g=1}^{|G|} \sum_{z=1}^{|Z|} \frac{\tilde{\pi}_g \cdot \prod_{j=1}^{J} \tilde{\psi}_{g,j,f^{u_{test}}}}{\tilde{\theta}_{g,z} \cdot \tilde{\phi}_{g,z,e} \cdot \tilde{\varphi}_{z,e,w}}, \tag{4.6}$$

where $\tilde{\pi}_g$, $\tilde{\psi}_{g,j,f^{u_{test}}}$, $\tilde{\theta}_{g,z}$, $\tilde{\phi}_{g,z,e}$, and $\tilde{\varphi}_{z,e,w}$ can be inferred according to equations above with the number of corresponding instances including the union of documents in D_{train} and document d_{test}.

To validate the effectiveness of the UGTE model, we conducted a series of experiments. We first investigated the impact of the number of user groups ($|G|$) on the performance of UGTE_ID that solely uses ID, in terms of both topic coherence and emotion classification. Figure 4.2 illustrates the topic coherence metrics (Coherence@10, Coherence@20, and Coherence@30) for various values of $|G|$, the number of user groups, and $|Z|$, the number of topics. When examining fewer topics ($|Z| = 25$ and $|Z| = 50$), we notice that topic coherence generally declines as the number of user groups grows, despite some initial variability. This observation implies that with a limited topic space, utilizing fewer user groups (i.e., broader groupings) favors the discovery of coherent topics. Conversely, with a larger number of topics ($|Z| = 100$), topic coherence remains relatively consistent across different $|G|$ values, suggesting that a larger topic space renders the model less affected by changes in the number of user groups.

Turning to emotion classification, Fig. 4.3 shows the accuracy and Cohen's Kappa score for different values of $|G|$ and $|Z|$. A similar trend emerges: Performance is generally better when $Z = 100$, and UGTE_ID performs stably across various group numbers. These findings suggest that the optimal number of user

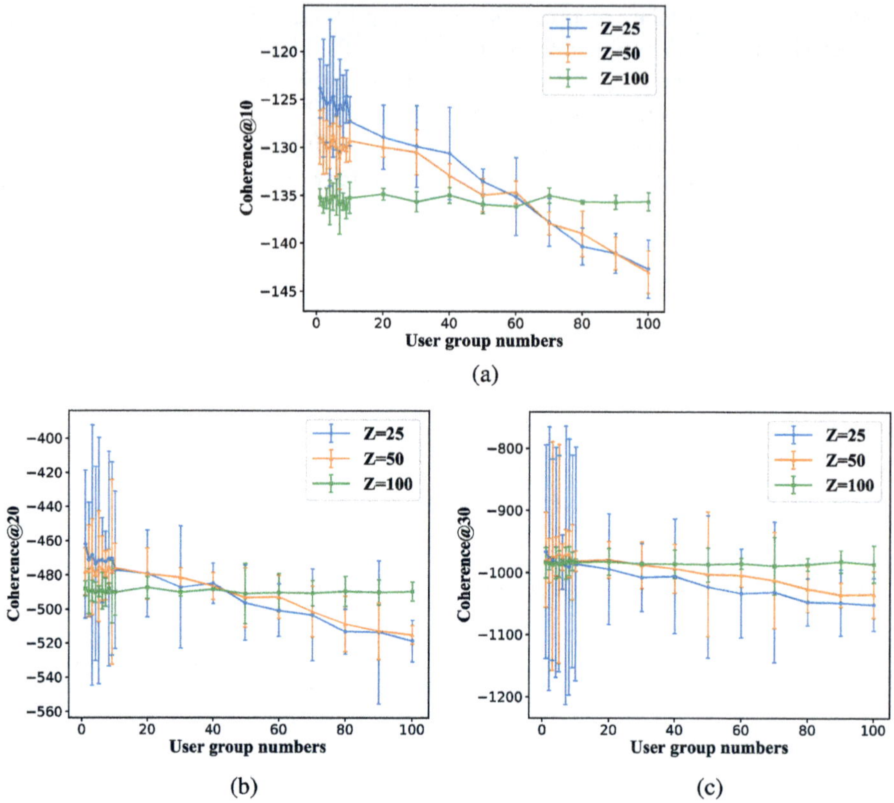

Fig. 4.2 Topic coherence along the number of user groups. (**a**) Coherence@10. (**b**) Coherence@20. (**c**) Coherence@30

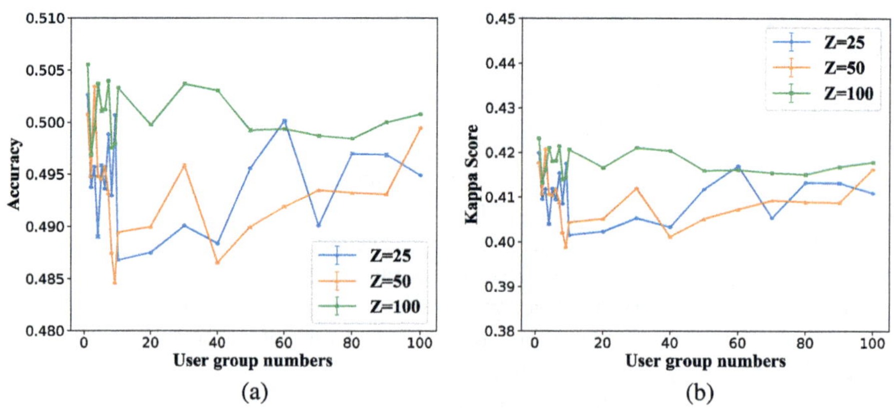

Fig. 4.3 Emotion classification performance along the number of user groups. (**a**) Accuracy. (**b**) Kappa score

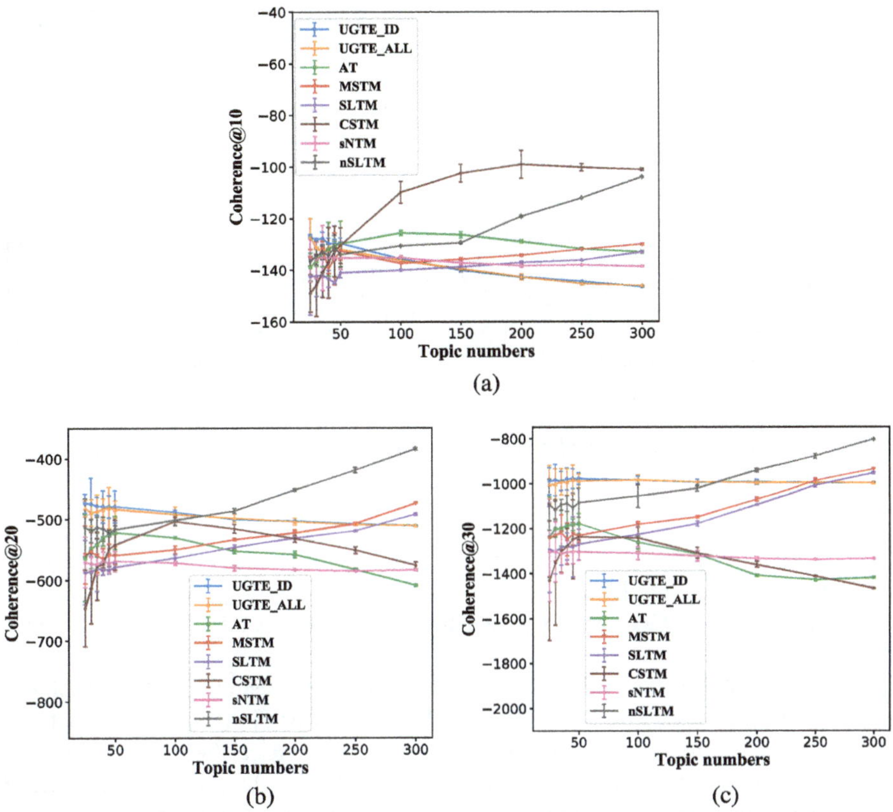

Fig. 4.4 Topic coherence comparison. (**a**) Coherence@10. (**b**) Coherence@20. (**c**) Coherence@30

groups depends on the complexity of the topic space. For simpler topic spaces, a smaller number of groups is preferable, while for more complex topic spaces, the model is more robust to the choice of $|G|$.

Next, we compared UGTE_ID and UGTE_ALL that uses all characteristic tags with several baseline models on the tasks of topic discovery and emotion classification. Figure 4.4 displays the topic coherence (Coherence@10, Coherence@20, and Coherence@30) for different values of $|Z|$ (with $|G|$ fixed at 10). Under Coherence@10, UGTE_ID outperforms the baselines when $|Z|$ is small. For Coherence@20 and Coherence@30, both UGTE_ALL and UGTE_ID perform better and exhibit more stability than other baselines. Notably, nSLTM demonstrates increasing coherence with larger $|Z|$, suggesting its suitability for mining numerous topics.

The accuracy and Cohen's Kappa scores for emotion classification are presented in Fig. 4.5. While nSLTM achieves the highest accuracy and Kappa score, UGTE_ID performs better than most other baselines. These results highlight

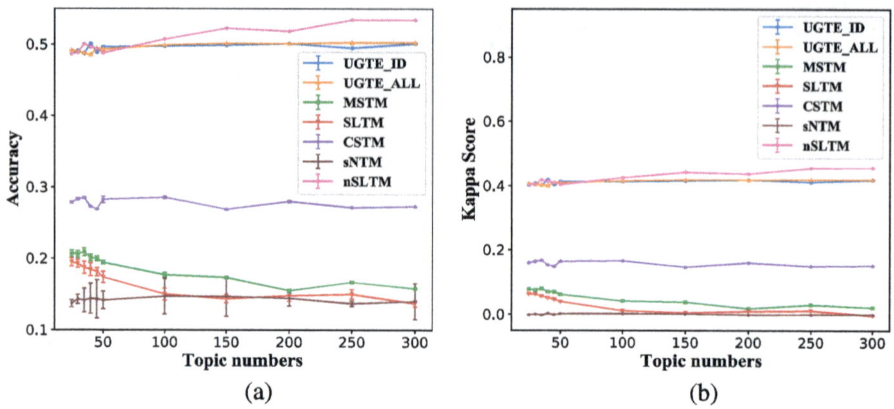

Fig. 4.5 Emotion classification performance comparison. (**a**) Accuracy over topics. (**b**) Kappa score over topics

UGTE's effectiveness in both topic discovery and emotion classification, especially when dealing with a smaller number of topics. It is worth noting that neural network-based models like nSLTM can excel when the task involves mining a large number of topics.

To further assess the model's ability to handle the feature sparsity problem inherent in short texts, we compared UGTE_ID and MSTM on the "short text" and "extremely short text" subsets of the ISEAR dataset. Figure 4.6 presents the topic coherence results. As expected, both UGTE_ID and MSTM achieve higher coherence on the "short text" subset compared to the "extremely short text" subset, confirming the increased difficulty posed by extremely short texts. However, UGTE_ID consistently outperforms MSTM on both subsets, indicating its superior ability to manage both short and extremely short texts and maintaining better stability.

Finally, Fig. 4.7 shows the emotion classification results on these subsets. UGTE_ID achieves significantly higher accuracy and Kappa scores than MSTM on both, with a particularly pronounced advantage on the "extremely short text" subset. This underscores UGTE's effectiveness in leveraging user characteristics to mitigate the feature sparsity problem.

These results strongly suggest that UGTE is particularly well suited for analyzing short and extremely short texts, where traditional topic-emotion models often struggle. To further demonstrate the effectiveness of our model in discovering meaningful user groups, Fig. 4.8 shows how different age groups exhibit varying emotional responses to the same topics, where UGTE_AGE and UGTE_NULL denote models that use solely AGE and no characteristic tags, respectively.

This model effectively addresses the feature sparsity problem by aggregating short texts within user groups while capturing the relationships between topics, emotions, and user characteristics at the group level. The introduction of user

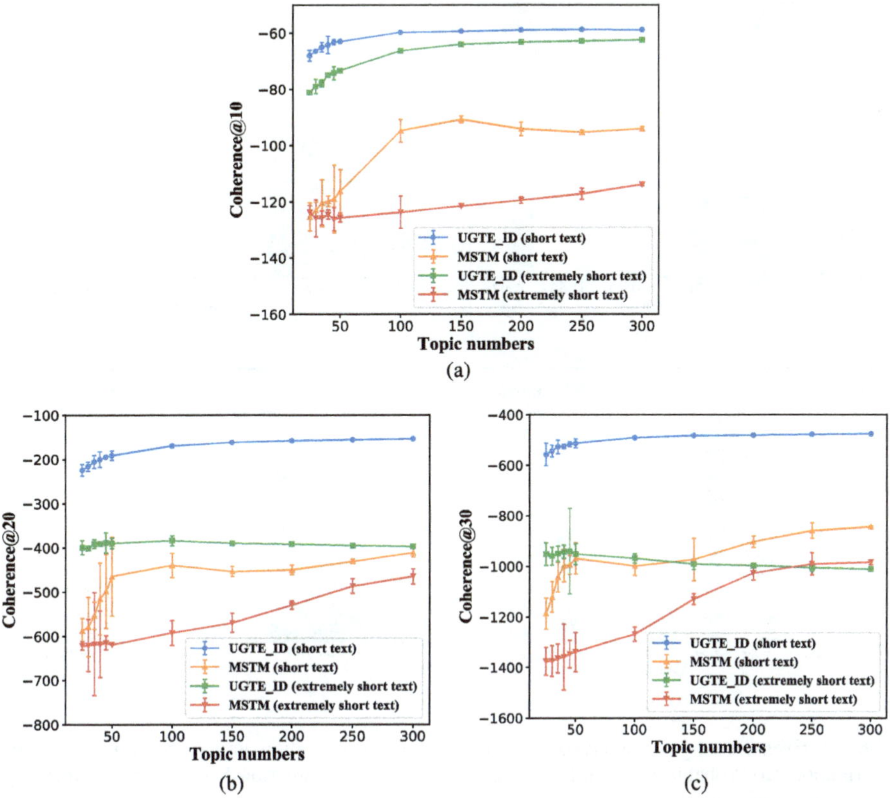

Fig. 4.6 Topic coherence on short and extremely short text. (**a**) Coherence@10. (**b**) Coherence@20. (**c**) Coherence@30

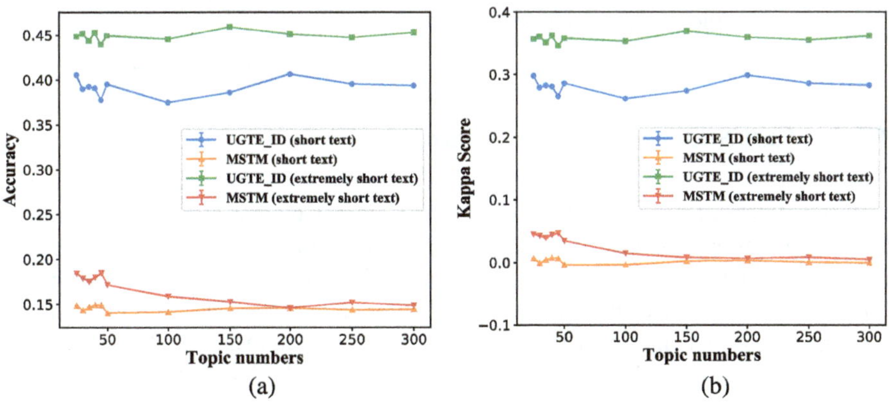

Fig. 4.7 Emotion classification performance on short and extremely short text. (**a**) Accuracy. (**b**) Kappa score

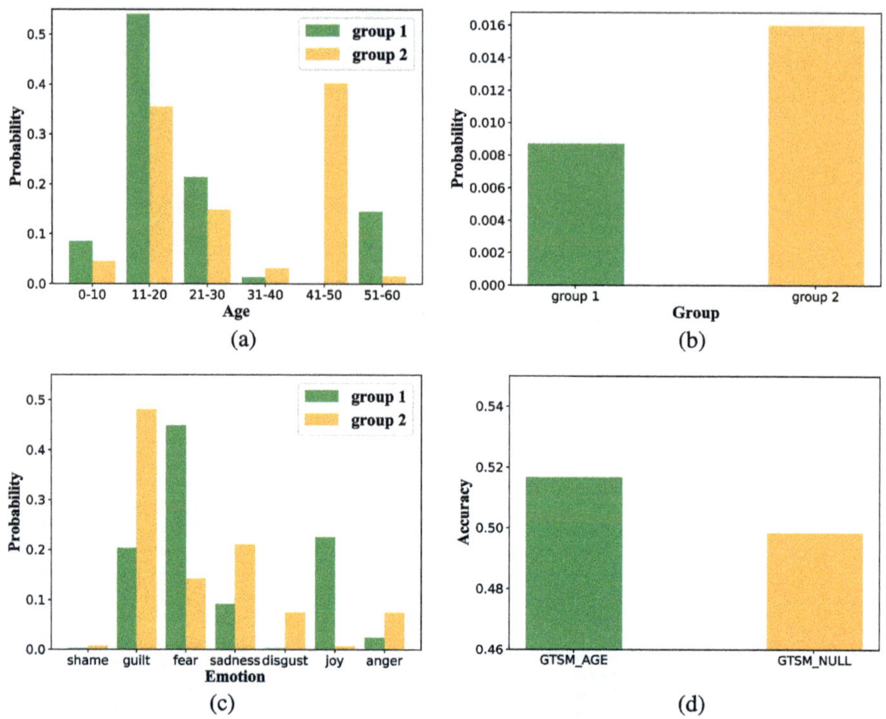

Fig. 4.8 Example of group analysis: emotion and topic distribution across age groups. (**a**) Age distribution. (**b**) Probability of "Intimate relationship" topic. (**c**) Emotion distribution. (**d**) Accuracy of emotion detection

characteristics helps identify semantic group structures and improve both emotion detection and topic discovery tasks.

4.1.2 *Fast Supervised Topic Models*

The prevalence of social media has generated massive short text data containing rich emotional expressions. However, detecting emotions from short texts remains challenging due to feature sparsity, as traditional topic models struggle to capture sufficient word co-occurrence patterns from limited context. This study introduces two supervised topic models with an accelerated algorithm to address this challenge.

Given a collection of documents $D = \{d_1, d_2, \ldots, d_{|D|}\}$ with vocabulary $W = \{w_1, w_2, \ldots, w_{|W|}\}$ and emotion labels $E = \{e_1, e_2, \ldots, e_{|E|}\}$, each document d_i is generated by user u_r and labeled with emotion e_k, denoted as $d_i^{r,k}$. The emotion annotations are represented by a real-valued matrix γ of size $N_D \times N_E$, where each row indicates a document's distribution over emotion labels.

4.1.2.1 Weighted Labeled Topic Model

The Weighted Labeled Topic Model (WLTM) introduces a one-to-many mapping between emotions and topics through a multiplier τ, which allows each emotion to be associated with multiple topics. As illustrated in Fig. 4.9, when $\tau = 5$, each emotion label is connected to five topics, enabling the model to capture a diverse range of emotional aspects. The generative process of WLTM begins by drawing $\phi_z \sim \text{Dirichlet}(\beta)$ for each topic z. Subsequently, for each X-term g_i, a topic indicator λ_{g_i} is generated based on the emotion labels. Following this, θ is drawn from a Dirichlet distribution conditioned on α and λ, i.e., $\theta \sim \text{Dirichlet}(\cdot|\alpha, \lambda)$. Finally, the model generates the topic assignment z_i and the corresponding words $\omega_{i,1}, \ldots, \omega_{i,X}$, completing the process. This structured approach ensures that the model effectively captures the nuanced relationships between emotions and topics. We present the graphical representation of WLTM in Fig. 4.10.

The posterior probability for topic inference is

$$P(z_i = z|z_{-i}, G) \propto \left(n_{-i,z} + \alpha\right) \times \frac{\gamma_{d_i,\lceil z/\tau \rceil}}{\sum_z \gamma_{d_i,\lceil z/\tau \rceil}} \times \prod_{x=1}^{X} \frac{n_{-i,\omega_{i,x}|z} + \beta}{n_{-i,\cdot|z} + N_\omega \beta}. \tag{4.7}$$

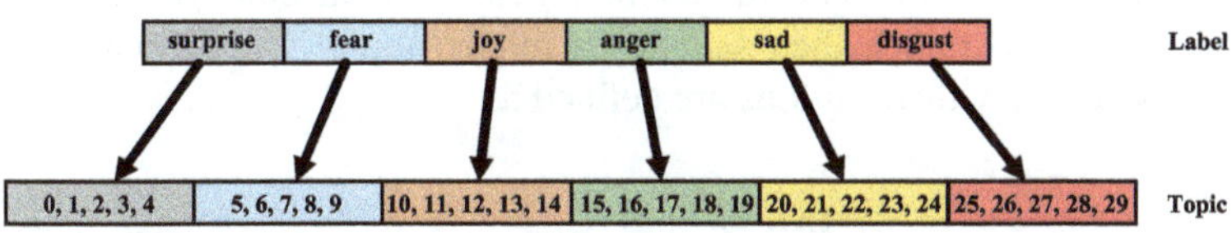

Fig. 4.9 Label-topic projection with $\tau = 5$

Fig. 4.10 Graphical representation of WLTM. Ψ is the prior for emotion labels of X-terms. τ is the number of topics per emotion. Λ is the topic presence/absence indicator. i denotes the ith X-term

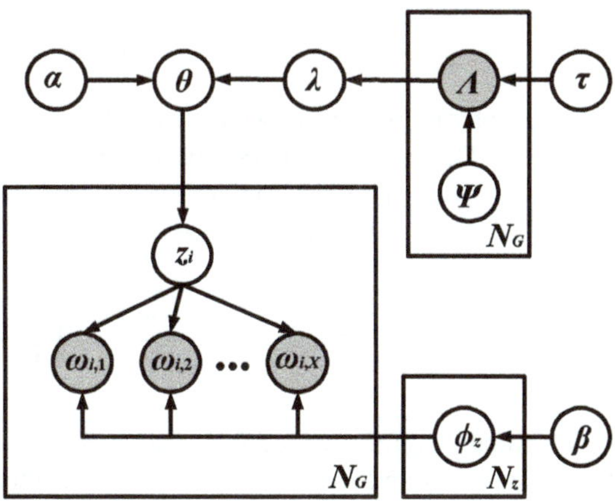

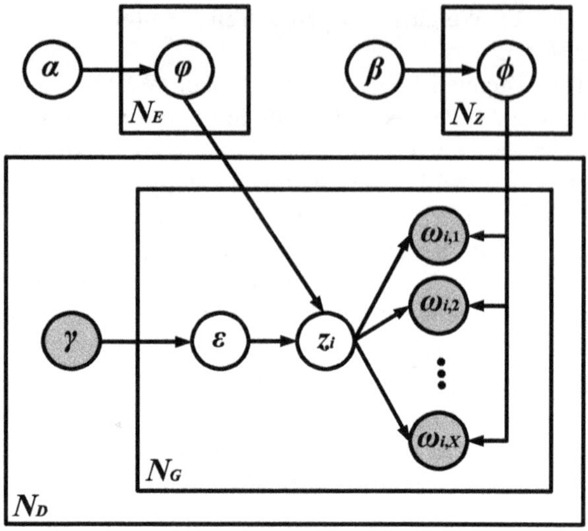

Fig. 4.11 Graphical representation of XETM

4.1.2.2 X-Term Emotion-Topic Model

The X-term Emotion-Topic Model (XETM) employs a hierarchical structure as illustrated in Fig. 4.11. The model first generates an emotion label ϵ under document-level constraints, then samples a related topic z, and finally generates word groups. The key distributions are defined as

$$\varphi_{\epsilon,z} = \frac{\alpha + n_{z|\epsilon}}{N_z\alpha + \sum_z n_{z|\epsilon}}, \tag{4.8}$$

$$\phi_{z,\omega} = \frac{\beta + n_{\omega|z}}{N_\omega\beta + \sum_\omega n_{\omega|z}}. \tag{4.9}$$

To accelerate inference, we combine the Alias method with Metropolis-Hastings sampling. This reduces the complexity from $O(N_{iter} \times N_G \times N_z)$ to $O(N_{iter} \times (N_\omega \times N_z + N_G))$ while maintaining model effectiveness. For topic sampling in WLTM, the acceptance probability is

$$\pi_{s\to t} = \frac{n_{-i,t} + \alpha}{n_{-i,s} + \alpha} \times \prod_{x=1}^{X} \frac{n_{-i,\omega_{i,x}|t} + \beta}{n_{-i,\omega_{i,x}|s} + \beta}. \tag{4.10}$$

The models effectively address feature sparsity by jointly modeling emotions and topics at the group level while achieving efficient inference through the accelerated algorithm. Experimental results on three datasets demonstrate superior performance in both accuracy and computational efficiency.

4.2 Topic-Aware Aspect Extraction

Here, we introduce our work entitled "TADC: A Topic-Aware Dynamic Convolutional Neural Network for Aspect Extraction" [16], which was developed for aspect extraction in fine-grained sentiment analysis, guided by document-topic distributions derived from neural topic models.

4.2.1 Aspect Extraction

Aspect extraction is a crucial task in fine-grained sentiment analysis, aimed at identifying explicit opinion targets from user-generated content. These opinion targets, or aspects, refer to the features or components of a product, service, or entity that users discuss or comment on. For instance, in a restaurant review, aspects might include "food," "service," or "ambiance." Identifying such aspects is key to understanding user sentiments and preferences, providing valuable insights for businesses to improve their products or services.

Formally, given a text $T = \{w_1, w_2, \ldots, w_n\}$, where T is a sequence of n words, the goal of aspect extraction is to identify a set of aspects $A = \{a_1, a_2, \ldots, a_m\}$, where a_i represents an aspect that is explicitly mentioned or implicitly inferred from the text.

Traditionally, aspect extraction has been approached using supervised learning techniques, often formulated as a sequence labeling problem. Recurrent Neural Networks (RNNs), known for their ability to capture sequential dependencies, have been widely used for this task. However, RNNs process information sequentially, which can be inefficient for large-scale datasets. By contrast, Convolutional Neural Networks (CNNs) enable parallel computation, but their fixed filters may struggle to effectively capture the unique characteristics of individual documents.

In this section, we introduce the proposed TADC model, which integrates topic modeling and dynamic convolution for aspect extraction. Specifically, the model utilizes a Variational Auto-Encoder (VAE) to generate an implicit mixture of topic representations. These topic representations are then leveraged by a dynamic convolutional neural network to perform sequence labeling for aspect term identification.

4.2.2 VAE-Based Topic Model

To investigate the inherent connection between topic modeling and aspect extraction, we introduce a VAE-based topic model, specifically designed to generate implicit topic representations. This topic model provides topic distributions that serve as essential features for the downstream aspect extraction task. The overall framework of the proposed approach is illustrated in Fig. 4.12.

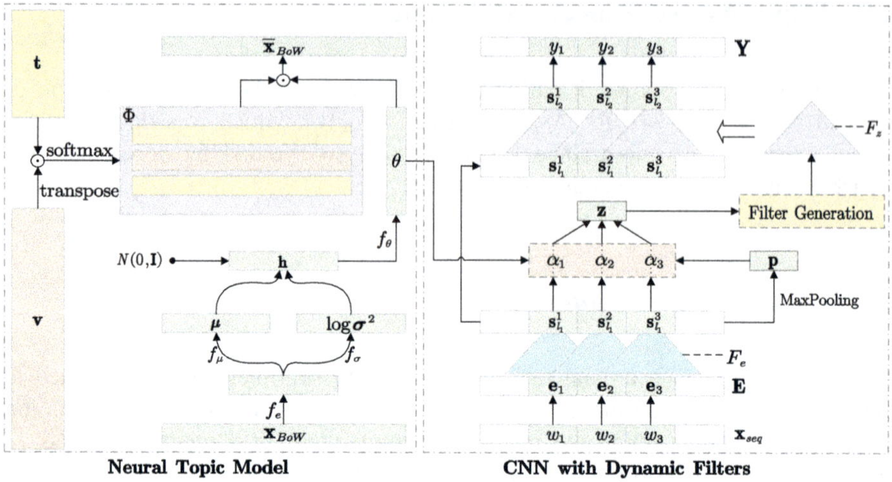

Fig. 4.12 Overall framework of TADC model. The model has two main parts: (1) VAE-based neural topic model and (2) DyCNN-based aspect extraction. F_e and F_z are filters for word and document embeddings

The model assumes that each document's topic distribution is defined based on a continuous latent variable $h \in \mathbb{R}^K$, where K is the number of topics. The process to generate a document representation $\tilde{\mathbf{x}}_{\text{BoW}}$ is defined as follows. The model represents each document's topic distribution using a continuous latent variable $h \in \mathbb{R}^K$, sampled from a Gaussian distribution $h \sim \mathcal{N}(\mu, \sigma^2)$. This latent variable captures the document's semantic structure. The topic mixture $\theta \in \mathbb{R}^K$ is then obtained by transforming h through a neural network f_θ, such that $\theta = f_\theta(h)$. The resulting θ is a valid probability distribution over topics, indicating the relevance of each topic to the document. Finally, the bag-of-words representation $\tilde{\mathbf{x}}_{\text{BoW}}$ is generated by combining θ with a learnable topic-word distribution matrix $\Phi \in \mathbb{R}^{K \times V}$, where V is the vocabulary size. This gives $\tilde{\mathbf{x}}_{\text{BoW}} = \theta \cdot \Phi$, reflecting both topic distributions and associated word probabilities.

Through this generative process, the model learns to extract meaningful topics while preserving flexibility and interpretability, making it highly suitable for applications such as aspect extraction.

To address the issue that Gaussian sampling $h \sim \mathcal{N}(\mu, \sigma^2)$ cannot be directly differentiated, the reparameterization trick is employed. Specifically, the variable h is expressed using a linear transformation of a standard Gaussian random variable $\epsilon \sim \mathcal{N}(0, 1)$. The transformation is given by $h = \mu + \epsilon \cdot \sigma$, where μ and σ are the learnable parameters. This formulation enables gradient-based optimization to update the model parameters while smoothing gradients during backpropagation.

For model inference, variational inference is applied to approximate the posterior distribution of the latent variable h. The objective is to minimize the Kullback-Leibler (KL) divergence between the approximate posterior $q(h|\mathbf{x})$ and the true posterior $p(h|\mathbf{x})$. The KL divergence is defined as

$$\begin{aligned} D_{KL}\left(q(h|x)\|p(h|x)\right) &= \int q(h|x)\log\frac{q(h|x)}{\frac{p(x|h)p(h)}{p(x)}}\,\mathrm{dh} \\ &= \log p(x)\int q(h|x)\,\mathrm{dh} + \int q(h|x)\log q(h|x)\,dh \\ &\quad - \int q(h|x)\log\left[p(x|h)p(h)\right]\mathrm{dh} \\ &= \log p(x)\cdot 1 + \int q(h|x)\log\frac{q(h|x)}{p(h)}\,\mathrm{dh} - \int q(h|x)\log p(x|h)\,dh \\ &= \log p(x) + D_{KL}\big(q(h|x)\|p(h)\big) - \mathbb{E}_{q(h|x)}\big[\log p(x|h)\big]. \end{aligned} \tag{4.11}$$

In Eq. (4.11), the first term denotes the marginal likelihood. The term can be obtained by omitting the integration $\int q(h|x)\,\mathrm{d}h$, which is equal to 1 in accordance with the definition of probability density. The second term along with the third term denotes the negative variational lower bound, based on which the loss function for NTM is defined. The terms can be directly obtained from the definitions of the KL divergence and mathematical expectation, respectively. Below, we give the expansion details of the second and third terms.

For $D_{\mathrm{KL}}(q(h|\mathbf{x})||p(h))$, $q(h|\mathbf{x})$ denotes the Gaussian distribution $\mathcal{N}(\mu,\sigma^2)$, where the latent variable h is drawn from, and $p(h)$ denotes a standard Gaussian prior $\mathcal{N}(0, 1)$. Therefore, the second term can be written as follows:

$$\begin{aligned} D_{KL}\left(q(h|x)\|p(h)\right) &= D_{KL}\left(\mathcal{N}(\mu,\sigma^2)\|\mathcal{N}(0,1)\right) \\ &= \int\frac{e^{-(h-\mu)^2/2\sigma^2}}{\sqrt{2\pi\sigma^2}}\log\frac{e^{-(h-\mu)^2/2\sigma^2}/\sqrt{2\pi\sigma^2}}{e^{-h^2/2}/\sqrt{2\pi}}\,\mathrm{dh} \\ &= \int\frac{e^{-(h-\mu)^2/2\sigma^2}}{\sqrt{2\pi\sigma^2}}\log\frac{e^{(h^2-(h-\mu)^2/\sigma^2)/2}}{\sqrt{\sigma^2}}\,dh \\ &= \frac{1}{2}\int\frac{e^{-(h-\mu)^2/2\sigma^2}}{\sqrt{2\pi\sigma^2}}\left(-\log\sigma^2 + h^2 - \frac{(h-\mu)^2}{\sigma^2}\right)dh \\ &= \frac{1}{2}\left(-\log\sigma^2 + \mu^2 + \sigma^2 - 1\right). \end{aligned} \tag{4.12}$$

For $-\mathbb{E}_{q(h|\mathbf{x})}\big[\log p(\mathbf{x}|h)\big]$, it denotes the reconstruction loss, which measures the discrepancy between the input bag-of-words representation $\mathbf{x}_{\text{BoW}}$ and the model's predicted distribution. The term is computed as the cross-entropy loss between the predicted output distribution $p(\tilde{\mathbf{x}}_{\text{BoW}}|h) = \theta \cdot \Phi$ and the input $\mathbf{x}_{\text{BoW}}$. The reconstruction loss can be expressed as

$$-\mathbb{E}_{q(h|\mathbf{x})}\big[\log p(\mathbf{x}|h)\big] = -\mathbf{x}_{\text{BoW}} \log p(\tilde{\mathbf{x}}_{\text{BoW}}|h). \tag{4.13}$$

Then the loss function for the Neural Topic Model (NTM) consists of two main parts. The first part, $\mathcal{J}$, combines the KL divergence and the reconstruction loss:

$$\mathcal{J} = \frac{1}{2}\sum_{i=1}^{K}\big(-\log(\sigma^{(i)})^2 + (\mu^{(i)})^2 + (\sigma^{(i)})^2 - 1\big) - \frac{1}{V}\sum_{i=1}^{V}\mathbf{x}_{\text{BoW}}^{(i)} \log p(\tilde{\mathbf{x}}_{\text{BoW}}^{(i)}|\mathbf{h}). \tag{4.14}$$

The model is expected to learn vector representations $\mathbf{t}$ as the most representative topics, but they may suffer from the redundancy problem during training. To ensure the uniqueness of each topic vector, we add a regularization term $\mathcal{R}$ to the loss function $\mathcal{J}$ as follows:

$$\mathcal{R} = \|\mathbf{t}_n \cdot \mathbf{t}_n^\top - \mathbf{I}\|, \tag{4.15}$$

where $\mathbf{t}_n$ is $\mathbf{t}$ with each row being normalized to 1 and $\mathbf{I}$ is an identity matrix. $\mathcal{R}$ reaches a minimum when the dot product between any two topic vectors is zero; therefore, the regularization term encourages orthogonality among the rows of $\mathbf{t}$ and penalizes redundancy between different topic vectors.

The final loss function of the NTM is obtained by

$$\mathcal{L}_{\text{NTM}} = \mathcal{J} + \gamma\mathcal{R}, \tag{4.16}$$

where γ is the trade-off parameter that controls the weight of the regularization term.

4.2.3 DyCNN-Based Aspect Extraction

The dynamic convolutional neural network (DyCNN) framework is designed for aspect extraction by incorporating dynamic filters that adapt to the input document.

Given an input document sequence $\mathbf{x}_{\text{seq}} = \{w_1, w_2, \ldots, w_N\}$ with N words, word embeddings are used to obtain their representations:

$$\mathbf{E} = \{\mathbf{e}_1, \mathbf{e}_2, \ldots, \mathbf{e}_N\} \in \mathbb{R}^{H\times N}, \tag{4.17}$$

where $\mathbf{e}_i \in \mathbb{R}^H$ is the H-dimensional embedding of word w_i. Then, a convolution operation followed by ReLU activation is applied on $\mathbf{E}$ to extract contextual features:

$$S_{l_1} = \text{ReLU}(F_e \otimes \mathbf{E}), \quad \mathbf{p} = \text{MaxPooling}(S_{l_1}), \tag{4.18}$$

where $F_e \in \mathbb{R}^{H \times \omega H}$ is a fixed filter matrix, and $\mathbf{p} \in \mathbb{R}^H$ represents the global context of the document.

To focus on words relevant to specific aspects, an attention mechanism generates a document embedding $\mathbf{z}$ as a weighted sum of contextual features:

$$\mathbf{z} = \sum_{i=1}^{N} \alpha_i \mathbf{s}_{l_1}^i, \quad \alpha_i = \frac{\exp\left(\mathbf{u}^\top \tanh\left(\mathbf{W}[\mathbf{s}_{l_1}^i; \mathbf{p}; \theta] + \mathbf{b}\right)\right)}{\sum_{j=1}^{N} \exp\left(\mathbf{u}^\top \tanh\left(\mathbf{W}[\mathbf{s}_{l_1}^j; \mathbf{p}; \theta] + \mathbf{b}\right)\right)}. \tag{4.19}$$

In the above, $\mathbf{u}$, $\mathbf{W}$, and $\mathbf{b}$ are trainable parameters, and θ is the topic distribution for the document generated from neural topic model.

To adapt the model to document-specific information, dynamic filters $\mathbf{F}_z$ are generated using the learned document embedding $\mathbf{z}$. These filters capture aspect-specific information:

$$\mathbf{F}_z = \mathbf{U} \cdot \text{diag}\left(\text{sigmoid}(\mathbf{z})\right) \cdot \mathbf{V} + \mathbf{B}, \tag{4.20}$$

where $\mathbf{U} \in \mathbb{R}^{H \times H}$, $\mathbf{V} \in \mathbb{R}^{H \times \omega H}$, and $\mathbf{B} \in \mathbb{R}^{H \times \omega H}$ are trainable weights.

The dynamic filters $\mathbf{F}_z$ are then applied to refine contextual features from the sequence:

$$\mathbf{S}_{l_2} = \text{ReLU}(\mathbf{F}_z \otimes \mathbf{S}_{l_1}), \tag{4.21}$$

where $\mathbf{S}_{l_2} \in \mathbb{R}^{H \times N}$ represents the refined aspect-aware features.

Deep neural networks are capable of integrating low-/mid-/high-level features in an end-to-end multilayer fashion, and network depth is of crucial importance. In this work, DyCNN can increase the network's depth by stacking multiple convolutional layers to learn hierarchical features. Figure 4.13 illustrates a two-block style of our method. For each block $b+1$, the context vector $\mathbf{p}_{b+1}$ used to compute the attention weights is updated based on the document embedding $\mathbf{z}_b$ and the context vector $\mathbf{p}_b$, which are both in the previous block, as follows:

$$\mathbf{p}_{b+1} = \mathbf{P} \cdot \mathbf{p}_b + \mathbf{z}_b, \tag{4.22}$$

where $\mathbf{P} \in \mathbb{R}^{H \times H}$ is a transformation matrix to be learned. Intuitively, $\mathbf{z}_b$ is dominated by the feature vectors with higher attention weights, and thus, $\mathbf{z}_b$ will approach the attended vectors corresponding to aspect words. The previously introduced topic mixture θ is incorporated into the attention mechanism for each block, guiding the model to pay more attention to the aspects.

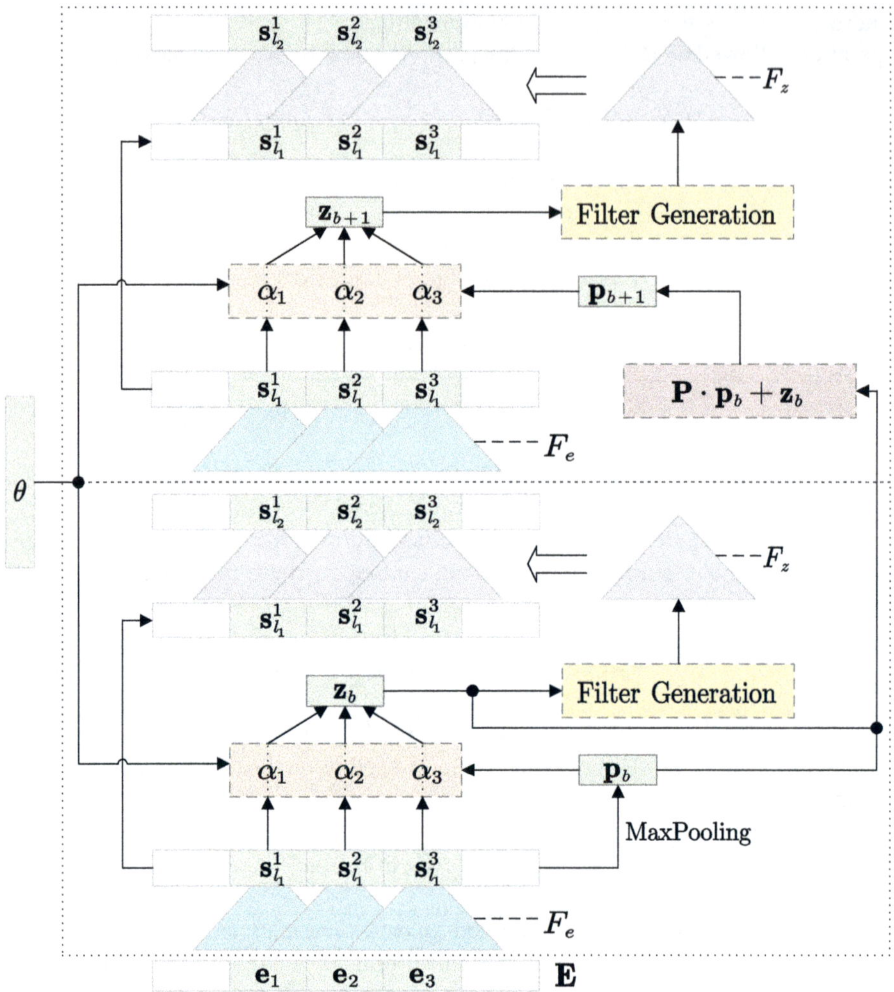

Fig. 4.13 Two-block style of our method

The output features $\mathbf{S}_{l_2} = \{\mathbf{s}_{l_2}^1, \mathbf{s}_{l_2}^2, \ldots, \mathbf{s}_{l_2}^N\} \in \mathbb{R}^{H \times N}$ of the top convolutional block are used as the final feature representations for the corresponding input word sequence $\mathbf{x}_{seq}$. Thus, the model predicts the aspect tag y_i for each word w_i using the refined features:

$$P(y_i|\mathbf{e}_i) = \text{softmax}(\mathbf{Q} \cdot \mathbf{s}_{l_2}^i), \tag{4.23}$$

where $\mathbf{Q} \in \mathbb{R}^{C \times H}$ is a trainable classification matrix, and $C = 3$ corresponds to $\{B, I, O\}$ tagging. The training loss is defined as the word-level cross-entropy:

$$\mathcal{L}_{\text{DyCNN}} = -\frac{1}{N}\sum_{i=1}^{N} P(y_i^*|\mathbf{e}_i)\log P(y_i|\mathbf{e}_i), \tag{4.24}$$

where y_i^* is the gold label for word w_i.

By combining context representation, attention-driven embeddings, and dynamic filters, DyCNN effectively captures aspect-specific information from documents. The overall model optimization is performed by minimizing this combined loss function, thus enabling the model to simultaneously learn holistic topic representations via the NTM module and refine aspect extraction through the DyCNN module.

The proposed TADC model integrates the two modules, i.e., the NTM module and the DyCNN module, which can be trained with their parameters updated simultaneously. We define the training objective of the entire framework as the linear combination of $\mathcal{L}_{NTM}$ and $\mathcal{L}_{DyCNN}$:

$$\mathcal{L} = \mathcal{L}_{NTM} + \lambda\mathcal{L}_{DyCNN}, \tag{4.25}$$

where λ is a hyperparameter that balances the effects of the two modules. By jointly optimizing both components, the model ensures that the topic-aware information guides the aspect-aware sequence labeling, leading to a more coherent and accurate representation for both tasks.

4.2.4 Experimental Results

Experiments are conducted on two benchmark datasets from SemEval challenges [8, 9]. The first dataset is composed of laptop reviews from subtask 1 of SemEval-2014 Task 4. The second dataset consists of restaurant reviews from subtask 1 (slot 2) of SemEval-2016 Task 5. In these datasets, aspect terms are labeled as spans of characters.

To evaluate the proposed model, the performance of the model is compared with those of three groups of baselines. The first group is single-task methods based on Conditional Random Field (CRF): CRF represents a basic CRF model that uses GloVe embeddings as input features. IHS_R&D [1] is a CRF model with rich lexical, syntactic, and statistical features, achieving the best performance in the laptop domain. NLANGP [11], based on CRFs, incorporates lexicon, syntactic, and cluster features, along with RNN outputs, excelling in the restaurant domain. WDEmb [15] enhances CRFs with word embeddings, linear context, and dependency path embeddings. The second group is single-task methods based on neural networks: LSTM [5] is a vanilla LSTM model utilizing pretrained word embeddings. CNN [14] is a four-layer CNN employing different filter sizes at its initial layers. BiLSTM-CNN-CRF [10] combines BiLSTM, CNN, and CRF layers for state-of-the-art named entity recognition. Seq2Seq4ATE [6] uses Seq2Seq with GRUs and position-aware attention for aspect extraction. The third group is based on multitask learning and exploits gold-standard opinion terms to improve model performance: RNCRF [12] combines CRFs with dependency-based recursive neural networks for joint aspect and opinion term extraction. CMLA [13] applies coupled multilayer attention for aspect-opinion co-extraction via tensor operators. MIN [4] employs LSTMs for aspect-opinion extraction and sentiment classification. HAST [3] leverages opinion summaries and extraction histories for aspect detection.

4.2.4.1 Main Result

Table 4.1 presents the results of different models on the aspect extraction task, where F_1 is employed as the evaluation metric. The values in the brackets denote the worst and the best results over five runs. The proposed TADC model achieves the best performance on both datasets. Specifically, it surpasses CNN with absolute gains of 2.96% and 4.36% and outperforms Seq2Seq4ATE, the strongest baseline, demonstrating its effectiveness.

TADC achieves these improvements without relying on additional resources, such as annotated opinion terms or sentiment lexicons. Its dynamic filters, incorporated with aspect information, enhance the model's ability to identify aspects accurately. Furthermore, TADC employs a topic-enhanced attention mechanism to capture document-level meaning and dependencies between adjacent words.

In addition to achieving better accuracy, TADC also benefits from faster training and inference due to its CNN-based architecture, which enables efficient parallelization. For instance, the training time of TADC is significantly lower compared to Seq2Seq4ATE, making it both effective and practical for real-world applications.

Table 4.1 F_1 performance comparison

Model	Laptop	Restaurant
CRF	74.01	69.56
IHS_R&D	74.55	–
NLANGP	–	72.34
WDEmb	75.16	–
LSTM	$75.20_{(73.69,75.81)}$	$71.07_{(70.69,71.79)}$
CNN	$78.22_{(77.60,79.19)}$	$71.38_{(71.00,72.30)}$
BiLSTM-CNN-CRF	$77.71_{(76.37,79.16)}$	$73.41_{(71.36,75.52)}$
Seq2Seq4ATE	$78.85_{(76.91,80.62)}$	$75.18_{(73.91,76.16)}$
RNCRF	78.42	69.72
CMLA	77.80	72.77
MIN	77.58	73.44
HAST	$77.83_{(77.48,78.48)}$	$71.69_{(70.87,72.30)}$
TADC	$\mathbf{81.18}_{(80.79,81.63)}$	$\mathbf{75.74}_{(75.43,76.00)}$

Table 4.2 Performance comparison in F_1 between TADC and its variants

Model	Laptop	Restaurant
CNN-vanilla	$78.72_{(78.48,79.03)}$	$71.91_{(71.76,71.97)}$
DyCNN	$79.31_{(79.06,79.60)}$	$74.30_{(73.74,74.57)}$
TADC	$\mathbf{81.18}*_{(80.79,81.63)}$	$\mathbf{75.74}*_{(75.43,76.00)}$

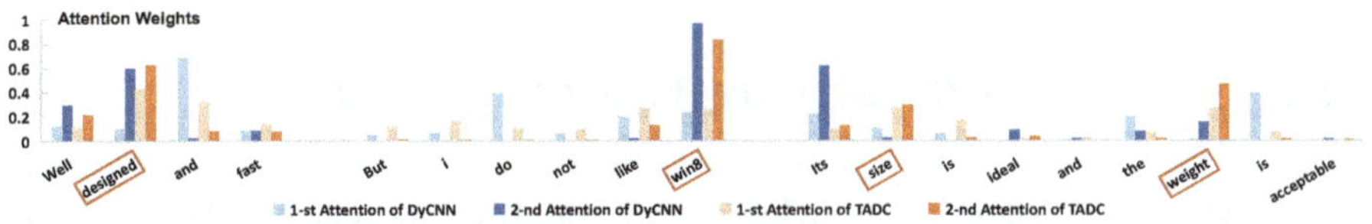

Fig. 4.14 Visualization of attention weights for different words in a sequence

4.2.4.2 Ablation Study

The following table and figure show the influence of the TADC model. There are two main variants of our model: (1) the DyCNN, which removes the NTM module from TADC, and (2) CNN-vanilla, which removes the attention mechanism and the filter generation network from the DyCNN. We report the results of TADC and its ablation baselines in Table 4.2. All values are averaged scores of five runs, and * indicates the result is statistically significant at the 0.05 level.

The DyCNN is 0.59% and 2.39% superior to CNN-vanilla on the laptop and restaurant datasets, respectively. This demonstrates that the dynamic filters generated from the input are more capable of exploring the aspect terms in the input. We can also see that TADC performs much better than the DyCNN. This implies that the topic information captured by the NTM can help guide the attention mechanism to pay more attention to the target aspects, and thus filters generated from the aspect information are more likely to attend to the aspect terms in the document.

4.2.4.3 Case Study

The actual attention weights for the word tokens of three sentences are visualized in Fig. 4.14. The vertical axis represents the normalized weights, where higher weights indicate larger probabilities of the corresponding words being part of an aspect term. Words highlighted within orange boxes correspond to the ground truth aspect terms. Specifically, a comparison is presented among the results from the single-block and two-block configurations, as well as those obtained using DyCNN and TADC.

For DyCNN, the attention mechanism in the first block (referred to as "first attention") generally does not effectively identify the significant components of the sentence. However, notable improvement is observed in the second block (referred to as "second attention"). Despite this improvement, certain errors persist when DyCNN is implemented with a two-block structure. For instance, in the third sentence, the model erroneously identifies "its" as an aspect term. This limitation of DyCNN is effectively addressed by TADC. As demonstrated in the results, the second attention mechanism of TADC is able to accurately identify all aspect terms while filtering out irrelevant components within the sentences. These findings validate that the topics captured by the Neural Topic Model facilitate attention toward target aspects that exhibit a strong correlation with specific topics.

4.2.5 Summary of Topic-Aware Aspect Extraction

In this section, we present TADC, a CNN-based model, to extract aspects from user-generated reviews. Since the filters employed by TADC are generated from the key aspect information inherent in the document, our model is capable of fully exploring the uniqueness of each document and hence effectively extracting the target aspects. Furthermore, owing to the strong correlations between topic modeling and aspect extraction, we exploit the topic information learned by an NTM to help identify aspects. The experimental results on two benchmark datasets demonstrate that TADC can not only achieve competitive performance in extracting aspects but can also produce coherent topics.

References

1. Chernyshevich M (2014) IHS R&D Belarus: cross-domain extraction of product features using CRF. In: Proceedings of the 8th international workshop on semantic evaluation, SemEval@COLING, pp 309–313
2. Feng J, Rao Y, Xie H, Wang FL, Li Q (2020) User group based emotion detection and topic discovery over short text. World Wide Web 23:1553–1587
3. Li X, Bing L, Li P, Lam W, Yang Z (2018) Aspect term extraction with history attention and selective transformation. In: Proceedings of the 27th international joint conference on artificial intelligence, pp 4194–4200
4. Li X, Lam W (2017) Deep multi-task learning for aspect term extraction with memory interaction. In: Proceedings of the 2017 conference on empirical methods in natural language processing, pp 2886–2892
5. Liu P, Joty SR, Meng HM (2015) Fine-grained opinion mining with recurrent neural networks and word embeddings. In: Proceedings of the 2015 conference on empirical methods in natural language processing, pp 1433–1443
6. Ma D, Li S, Wu F, Xie X, Wang H (2019) Exploring sequence-to-sequence learning in aspect term extraction. In: Proceedings of the 57th conference of the association for computational linguistics, pp 3538–3547
7. Pang J, Rao Y, Xie H, Wang X, Wang FL, Wong TL, Li Q (2021) Fast supervised topic models for short text emotion detection. IEEE Trans Cybern 51(2):815–828
8. Pontiki M, Galanis D, Papageorgiou H, Androutsopoulos I, Manandhar S, Al-Smadi M, Al-Ayyoub M, Zhao Y, Qin B, Clercq OD, Hoste V, Apidianaki M, Tannier X, Loukachevitch NV, Kotelnikov EV, Bel N, Jiménez-Zafra SM, Eryigit G (2016) SemEval-2016 task 5: aspect based sentiment analysis. In: Proceedings of the 10th international workshop on semantic evaluation, SemEval@NAACL-HLT, pp 19–30
9. Pontiki M, Galanis D, Pavlopoulos J, Papageorgiou H, Androutsopoulos I, Manandhar S (2014) Semeval-2014 task 4: aspect based sentiment analysis. In: Proceedings of the 8th international workshop on semantic evaluation, SemEval@COLING, pp 27–35

10. Reimers N, Gurevych I (2017) Reporting score distributions makes a difference: performance study of LSTM-networks for sequence tagging. In: Proceedings of the 2017 conference on empirical methods in natural language processing, pp 338–348
11. Toh Z, Su J (2016) NLANGP at semeval-2016 task 5: improving aspect based sentiment analysis using neural network features. In: Proceedings of the 10th international workshop on semantic evaluation, SemEval@NAACL-HLT, pp 282–288
12. Wang W, Pan SJ, Dahlmeier D, Xiao X (2016) Recursive neural conditional random fields for aspect-based sentiment analysis. In: Proceedings of the 2016 conference on empirical methods in natural language processing, pp 616–626
13. Wang W, Pan SJ, Dahlmeier D, Xiao X (2017) Coupled multi-layer attentions for co-extraction of aspect and opinion terms. In: Proceedings of the 31st AAAI conference on artificial intelligence, pp 3316–3322
14. Xu H, Liu B, Shu L, Yu PS (2018) Double embeddings and CNN-based sequence labeling for aspect extraction. In: Proceedings of the 56th annual meeting of the association for computational linguistics, pp 592–598
15. Yin Y, Wei F, Dong L, Xu K, Zhang M, Zhou M (2016) Unsupervised word and dependency path embeddings for aspect term extraction. In: Proceedings of the 25th international joint conference on artificial intelligence, pp 2979–2985
16. Zhang Z, Rao Y, Lai H, Wang J, Yin J (2021) TADC: a topic-aware dynamic convolutional neural network for aspect extraction. IEEE Trans Neural Netw Learn Syst 34(8):3912–3924

Chapter 5
Discussions

Abstract In this chapter, we will first summarize the content of this book, including classical and modern topic models, as well as how these models can be applied to solve practical tasks of emotion detection and aspect extraction. Then, we will point out the following promising directions for further investigation. (1) The enhancement of evaluation criteria: In evaluating flat topic models, the presence of polysemous words can significantly impact the coherence and diversity of topics, yet their influence is not fully captured by existing criteria. In evaluating hierarchical topic models, there is a notable lack of automated metrics specifically targeting the parent-child relationships within the topic hierarchy. (2) The integration of prior knowledge: Future work could explore integrating prior knowledge to guide the discovery process of topic structures, so as to improve both the interpretability and domain-specific applicability of the models. (3) The expansion of model capabilities: Considering the proliferation of multilingual and cross-domain data, it is valuable to construct a unified topic model capable of handling diverse text inputs. Finally, we expect more application scenarios to enhance the practical impact of topic models.

5.1 Conclusion

This book provides an exhaustive exploration of the evolving landscape of topic modeling, from foundational concepts and traditional methods to the latest advancements that significantly enhance text data mining and analysis. We begin by introducing the basics of topic models and reviewing seminal works, including flat topic models and neural topic models. The discussion then delves into hierarchical topic models, exploring advanced approaches such as nonparametric tree-structured models, forest-structured models, and Gaussian Mixture Distribution Hierarchical Topic Models. These models not only uncover deep semantic structures within document collections but also capture the hierarchical relationships between different topics, which is crucial for a deeper understanding of textual information.

Y. Rao, Q. Li, *Topic Modeling*, Machine Learning: Foundations, Methodologies, and Applications, https://doi.org/10.1007/978-981-96-8853-1_5

Furthermore, the chapter on modern topic models explores cutting-edge methodologies that significantly advance the field beyond traditional approaches. By integrating advanced machine learning techniques such as neural networks, nonparametric clustering, and contrastive learning, these models address key challenges like semantic alignment over time, noise in social media data, and the dynamic updating of topic structures. Innovations like anchor word clustering, label aligned priors, and geometric representations in box embedding spaces enhance the interpretability and coherence of discovered topics, providing more flexible and meaningful hierarchical structures. Collectively, these advancements pave the way for robust and versatile applications of topic modeling across diverse domains, setting a new standard for uncovering deeper insights from textual data.

We have demonstrated how these models can be applied to solve practical problems, such as continuous learning environments for topic discovery, emotion detection, and aspect extraction. Experimental results highlight the superior performance of our methods across various evaluation metrics, effectively mining rational and unique topic structures while ensuring rapid convergence.

In summary, this book offers a comprehensive theoretical framework and technical toolkit for researchers and practitioners, promoting a deeper understanding and processing of textual information. By introducing innovative methodologies and technologies, we emphasize the ongoing evolution in the field of topic modeling, laying a solid foundation for future research. Overall, with continuously growing computational power and evolving technological means, the field of topic modeling stands at the threshold of numerous opportunities and challenges. We look forward to subsequent research breaking new ground and delivering richer, more valuable knowledge outcomes for society.

5.2 Future Work

Despite the extensive coverage and effective solutions presented in this book, the field of topic modeling still presents numerous unresolved challenges and exciting opportunities. Below are several promising directions for further investigation.

Enhancement of Evaluation Criteria Currently, the evaluation standards used to assess the quality of topics generated by different models still have certain limitations. Existing methods often overlook critical aspects that are essential for a comprehensive evaluation.

In flat topic models, the diversity of topics concerning polysemous words (i.e., words with multiple meanings) is frequently neglected. Current metrics do not adequately account for this complexity, potentially leading to an oversimplified view of topic diversity. The presence of polysemous words can significantly impact the coherence and distinctiveness of topics, yet their influence is not fully captured by existing evaluation criteria.

For hierarchical topic models, there is a notable lack of automated evaluations specifically targeting the parent-child relationships within the topic hierarchy. Assessing these relationships is crucial because they define the semantic connections between general and specific themes, which are fundamental to understanding the rationality and interpretability of the structure. Without proper assessment of these relationships, it is challenging to ensure that the hierarchical organization truly reflects the underlying semantics of the data.

Furthermore, manual evaluations such as topic intrusion and word intrusion tests, while providing valuable insights, require significant human resources and introduce considerable subjectivity. The reliance on human judgment can lead to inconsistencies and may not scale well for large-scale analyses. These subjective assessments can also be time-consuming and costly, limiting their practical applicability.

To address these limitations, future work should focus on developing more robust and automated evaluation metrics to (i) account for the diversity of topics in flat models, especially considering polysemous words, to provide a more accurate representation of topic coherence and distinctiveness, (ii) specifically evaluate the parent-child relationships in hierarchical models to ensure meaningful and coherent topic hierarchies, thereby enhancing the interpretability of the model's output, and (iii) reduce the reliance on labor-intensive and subjective human evaluations, promoting scalable and objective assessments that can handle large datasets efficiently.

By addressing these gaps, new evaluation criteria will offer a more precise and nuanced understanding of topic model performance, ultimately enhancing their reliability and applicability in real-world scenarios. This advancement is crucial for driving forward the field of topic modeling and ensuring that models can effectively uncover deeper insights from textual data.

Integration of Prior Knowledge Most current topic models primarily rely on intrinsic data features for training, with limited consideration of external knowledge sources. Future work could explore integrating prior knowledge from general knowledge graphs like WordNet[1] or ConceptNet[2] to guide the discovery process of topic structures. This approach can enhance both the interpretability and domain-specific applicability of the models.

Expansion of Model Capabilities Present-day topic models largely focus on static text data, whereas real-world text data tends to be dynamic. Future models should adapt to these changes, enabling effective tracking and updating of topics over time. Moreover, considering the proliferation of multilingual and cross-domain data, constructing a unified framework capable of handling diverse text inputs will be an important research direction.

Deepening Application Scenarios Although this book showcases the potential of topic models in areas like emotion detection and aspect extraction, many other

---

[1] https://wordnet.princeton.edu/

[2] https://conceptnet.io/

application scenarios remain unexplored, for instance, utilizing user behavior logs as input in personalized recommendation systems to predict points of interest or combining social network analysis to investigate the formation mechanisms and evolutionary patterns of public opinion. Such applications promise to broaden the practical impact and societal value of topic models.

GPSR Compliance
The European Union's (EU) General Product Safety Regulation (GPSR) is a set of rules that requires consumer products to be safe and our obligations to ensure this.

If you have any concerns about our products, you can contact us on

ProductSafety@springernature.com

In case Publisher is established outside the EU, the EU authorized representative is:

Springer Nature Customer Service Center GmbH
Europaplatz 3
69115 Heidelberg, Germany

www.ingramcontent.com/pod-product-compliance
Lightning Source LLC
Chambersburg PA
CBHW070401030126
37625CB00052B/54
* 9 7 8 9 8 1 9 6 8 8 5 2 4 *